Functional Testing
in
Human
Performance

Michael P. Reiman

Wichita State University

Robert C. Manske

Wichita State University

Human Kinetics

Library of Congress Cataloging-in-Publication Data

Reiman, Michael P., 1965-
   Functional testing in human performance / Michael P. Reiman, Robert C. Manske.
     p. ; cm.
   Includes bibliographical references and index.
   ISBN-13: 9-780-7360-6879-6 (hardcover)
   ISBN-10: 0-7360-6879-1 (hardcover)
   1. Exercise tests. 2. Function tests (Medicine) 3. Disability evaluation. I. Manske,
Robert C. II. Title.
   [DNLM: 1. Physical Fitness--physiology. 2. Physical Therapy Modalities. 3. Athletic
Injuries--diagnosis. 4. Exercise Test. QT 255 R363f 2009]
   RM725.R38 2009
   615.8'2--dc22

                2008048271

ISBN-10: 0-7360-6879-1
ISBN-13: 9-780-7360-6879-6

The Web addresses cited in this text were current as of October 28, 2008, unless otherwise noted.

Acquisitions Editor: Loarn D. Robertson, PhD; **Developmental Editor:** Maggie Schwarzentraub; **Assistant Editor:** Nicole Gleeson; **Copyeditor:** Joyce Sexton; **Proofreader:** Kathy Bennett; **Indexer:** Craig Brown; **Permission Manager:** Dalene Reeder; **Graphic Designer:** Bob Reuther; **Graphic Artist:** Dawn Sills; **Cover Designer:** Keith Blomberg; **Photographer (cover):** Neil Bernstein; **Photographer (interior):** Neil Bernstein; **Visual Production Assistant:** Joyce Brumfield; **Photo Office Assistant:** Jason Allen; **Art Manager:** Kelly Hendren; **Associate Art Manager:** Alan L. Wilborn; **Illustrator:** Tammy Page; **Printer:** Sheridan Books

We thank Player Development Solutions in Wichita, Kansas, for assistance in providing the location for the photo shoot for this book.

Printed in the United States of America   10 9 8 7 6 5 4

The paper in this book is certified under a sustainable forestry program.

Human Kinetics
Web site: www.HumanKinetics.com

United States: Human Kinetics
P.O. Box 5076
Champaign, IL 61825-5076
800-747-4457
e-mail: humank@hkusa.com

Canada: Human Kinetics
475 Devonshire Road, Unit 100
Windsor, ON N8Y 2L5
800-465-7301 (in Canada only)
e-mail: info@hkcanada.com

Europe: Human Kinetics
107 Bradford Road
Stanningley
Leeds LS28 6AT, United Kingdom
+44 (0)113 255 5665
e-mail: hk@hkeurope.com

Australia: Human Kinetics
57A Price Avenue
Lower Mitcham, South Australia 5062
08 8372 0999
e-mail: info@hkaustralia.com

New Zealand: Human Kinetics
P.O. Box 80
Torrens Park, South Australia 5062
0800 222 062
e-mail: info@hknewzealand.com

To my wife, Kim, and children, Carly and Seth,
for their endearing support and understanding. Without
such support and understanding, projects such as this
would not be accomplished and would mean nothing.

Michael P. Reiman

To my wife, Julie, and my children, Rachael, Halle, and Tyler,
for their constant support with all of my various projects.

Robert C. Manske

Contents

Test Finder **vii** ■ Preface **xiii** ■ Acknowledgments **xvii**

PART I **Basics of Functional Testing** 1

1 Essential Concepts and Terms 3

2 Test Administration 9

3 Integration of Functional Testing Into Everyday Practice 17

PART II **Testing Procedures and Protocols for Discrete Physical Parameters** 29

4 Anthropometric Assessment 31

5 Muscle Length Assessment 39

6 Fundamental Movement Testing 85

7 Balance Testing 103

8 Aerobic Testing 119

9 Strength and Power Testing 131

10 Speed, Agility, and Quickness Testing 191

PART III — Testing Procedures and Protocols for Regional Physical Parameters 209

11 Trunk Testing 211

12 Upper Extremity Testing 241

13 Lower Extremity Anaerobic Power Testing 263

Appendix—Reproducible Forms **275** ■ References **285**

Index **299** ■ About the Authors **307** ■ DVD Menu and User Instructions **310**

Test Finder

Test	Page number	Included on DVD?
CHAPTER 4 ANTHROPOMETRIC ASSESSMENT		
Body Mass Index Assessment	33	
Girth Assessments	32	
Pelvic Girth Assessment	35	
Torso Height Assessment	34	
Waist-to-Hip Ratio Assessment	33	
CHAPTER 5 MUSCLE LENGTH ASSESSMENT		
Biceps Assessment	53	
External Rotators of the Hip Assessment	60	
Gastrocnemius Assessment	75	
Hamstring Length: Active Supine 90/90 Position Assessment	72	
Hamstring Length: Passive Supine 90/90 Position Assessment	74	
Hamstrings: Passive Straight Leg Raise Method	73	
Hip Adductor Assessment (Long Versus Short)	67	
Hip Extensors Assessment	58	
Hip Flexibility: Figure Four Test	71	
Hip Flexors Assessment	59	
Internal Rotators of the Hip Assessment	61	
Latissimus Dorsi Assessment	46	
Latissimus Dorsi Assessment	57	
Levator Scapulae Assessment	41	
Lower (Sternal)—Pectoralis Major Assessment	49	
Lumbar Erector Spinae Assessment	54	
Ober's Test—Iliotibial Band Assessment	68	
Pectoralis Major Assessment	47	
Pectoralis Minor Assessment	50	
Pectoralis Minor Assessment (Borstad Method)	51	
Piriformis (FAIR Test)	70	

(continued)

TEST FINDER *(continued from previous page)*

Test	Page number	Included on DVD?
Piriformis Assessment	69	
Psoas Assessment I	62	
Psoas Assessment II	63	
Quadratus Lumborum Assessment I	55	
Quadratus Lumborum Assessment II	56	
Rectus Femoris Assessment	64	
Soleus Muscle Length Test: Supine	76	
Sternocleidomastoid Assessment	44	
Suboccipital Muscles Assessment	45	
Tensor Fasciae Latae Assessment	66	
Triceps Assessment	52	
Upper (Clavicular)—Pectoralis Major Assessment	48	
Upper Trapezius Assessment—Sitting	42	
Upper Trapezius Assessment—Supine	43	
CHAPTER 6 FUNDAMENTAL MOVEMENT TESTING		
Active Straight Leg Raise Assessment	97	X
Cervical Deep Flexor Muscle Assessment	86	
Deep Squat Assessment	91	X
Hip Abduction Assessment	88	X
Hip Extension Assessment	87	X
Hurdle Step Assessment	93	X
In-Line Lunge Assessment	95	X
Rotational Stability Assessment	99	X
Shoulder Mobility Assessment	96	X
Trunk Flexion Assessment	89	X
Trunk Stability Push-Up Assessment	98	
CHAPTER 7 BALANCE TESTING		
Balance Error Scoring System	114	
Four Square Step Test	106	
Functional Reach Test	110	X
Multiple Single-Leg Hop Stabilization Test	115	X
Romberg Test	111	

Test	Page number	Included on DVD?
Side-Step Test	107	
Single-Leg Stance Test	104	
Star Excursion Balance Test	108	X
Static Balance: Stork Test	105	
Tandem Walking	112	
Tinetti Test	113	
CHAPTER 8 AEROBIC TESTING		
1.5-Mile Run Test	125	
12-Minute Run Test	125	
20-Meter Shuttle Run Test	123	
Chester Step Test	122	
Multistage Fitness Test (20-Meter Shuttle Run, Yo-Yo Test)	124	
One-Mile Walk Test	120	
Rockport Walk Test	121	
CHAPTER 9 STRENGTH AND POWER TESTING		
1RM Back Squat	141	
1RM Bench Press	144	
1RM Leg Press	143	
Alternate Single-Leg Squat Test	147	X
Carioca Drill or Test	134	
Co-contraction Test	132	
Figure 8 Hop Test	160	X
Hexagon Test (Bilateral Lower Extremity Jump)	164	X
Hop Testing After Fatigue	166	
Knee Bending in 30 Seconds	137	
Lunge Test	135	
Maximal Controlled Leap	157	
Modified Hexagon Hop Test	165	X
One-Legged Cyclic Hop Test	138	
Side-Hop Test	162	
Single-Hop Test	163	

(continued)

Test	Page number	Included on DVD?
Single-Leg Crossover Hop for Distance Test	156	X
Single-Leg Hop for Distance	149	
Single-Leg Inclined Squat Test (20 Reps and 50 Reps)	139	X
Single-Leg Squat	146	
Six-Meter Timed Hop	154	
Stair Hopple Test	158	
Standing Long Jump	148	
Step-Down	133	
Triple Hop for Distance	155	
Triple Jump for Distance Test	153	X
Up–Down Test	161	
Vertical Jump Test	151	X
CHAPTER 10 SPEED, AGILITY, AND QUICKNESS TESTING		
505 Agility Test	194	X
Backward Movement Agility Test	200	
Edgren Side-Step Test	195	X
Hurdle Test	198	
Illinois Agility Test	199	X
Pro Agility (5-10-5) Test	193	X
Slalom Test	197	
Three-Cone Drill Test	196	X
T-Test	192	X
Zigzag Run Test	201	X
CHAPTER 11 TRUNK TESTING		
Abdominal Dynamic Endurance	226	
Abdominal Stage Test	219	X
Deep Neck Flexor Test	213	X
Double-Leg Lowering Test	217	X
Endurance of Abdominal Muscles Test	224	X
Endurance of Extensors	222	
Endurance of Lateral Flexors (Side Bridge)	221	
Extensor Dynamic Endurance Test	227	X

Test	Page number	Included on DVD?
Loaded Forward Reach	230	
Progressive Isoinertial Lifting Evaluation (PILE)	231	
Prone Bridge	228	
Repetitive Box-Lifting Task (RBLT)	232	
Segmental Multifidus Test	215	X
Sit-Up Endurance Testing	220	
Supine Bridge	229	
Trunk Curl-Up Test	216	X
CHAPTER 12 UPPER EXTREMITY TESTING		
Alternate Pull-Up Test	244	X
Backward Overhead Medicine Ball Throw Test	247	X
Closed Kinetic Chain Upper Extremity Test	253	X
Flexed Arm Hang	242	
Functional Throwing Performance Index (FTPI)	254	
Medicine Ball Toss	246	
Pull-Up Test	243	
Push-Up Test	245	
Seated Chest Pass	250	
Seated Shot-Put Throw	251	
Sidearm Medicine Ball Throw	249	
Underkoffler Softball Throw for Distance	245	
CHAPTER 13 LOWER EXTREMITY ANAEROBIC POWER TESTING		
300 Shuttle Run	264	
300-Meter Sprint	270	
Bosco Test	267	
Lower Extremity Functional Test	265	X
Plyometric Leap Test	269	X
Running-Based Anaerobic Sprint Test (RAST)	264	
Squat Jump Test	268	X
Wingate Anaerobic Power Test	266	

Preface

Functional performance is an aspect of everyday life, whether for the elite athlete or the industrial worker. The commonality among all athletes, students, workers, or any other group is that ability in one or more aspects of physical function is needed in order for them to be successful at their skills. Successful performance in one's daily tasks often requires several physical components, and these components can be difficult to assess and quantify functionally. A number of components of performance have been investigated, although the research is reported in many different types of publications and by researchers in various disciplines, including the allied health disciplines.

Individual physical components of performance have been studied extensively in the exercise physiology domain. Strength and power components of physical performance have also been studied in great depth by physical therapists, athletic trainers, exercise physiologists, and strength and conditioning professionals. Other physical components such as flexibility and movement skills have also been investigated in multiple disciplines.

The distinction between physical and functional performance testing is important in our context. We believe that physical performance testing addresses attributes commonly tested in a clinical, athletic, or work setting. These include attributes such as physical strength, balance, proprioception, endurance, and muscle flexibility. We have chosen to define functional testing as the use of a variety of physical skills and tests to determine (1) one's ability to participate at the desired level in sport, an occupation, or recreation or to return to participation in a safe and timely manner without functional limitations and (2) one's ability to move through up to three planes of movement as assessed via nontraditional testing that provides qualitative and quantitative information related to specialized motions involved in sport, exercise, and occupations. The tests that constitute functional testing can include those covered in this book: assessments of flexibility; functional movement patterns; balance;

aerobic capabilities; muscular strength, power, and endurance; and speed and agility.

Obviously there are components common to both physical and functional testing. Functional testing is intended as an assessment of the client's true functional ability, and that functional ability can include strength and power, for example. In other words, physical attributes are a part of functional performance. Thus in some cases it is not easy to draw the line between physical testing and functional testing. In many cases, though, the two types of testing are distinct. Isokinetic testing of lower extremity quadriceps and hamstring strength is a physical test of strength, endurance, or power depending on the goals of the testing; but an isokinetic test is not a functional test. It does not determine a person's ability to squat on one leg to pick keys up off the ground or lower his or her body weight from an 18 in. step in a factory. The strength of the quadriceps and hamstrings as measured isokinetically definitely contributes to the ability to perform these tasks; but so do multiple other factors, including hip, knee, and ankle mobility; flexibility; and strength and lower extremity balance. The clinician would undoubtedly assess each of these components separately and may find that all the results are normal, but the client may still have limitations in functional ability. While it is imperative to assess each of these components prior to functional testing, these components do not always correlate highly with function. The ultimate functional test is a test in which the client performs the task. Having the individual client actually step off an 18 in. step enables the clinician to qualitatively and quantitatively measure his or her functional ability to perform this task. Often this is what really matters to the client. Functional testing presents multiple problems, most importantly the problem of how to quantify the results. There needs to be a balance between physical testing and the actual performance of the nonmeasurable task. Finding this balance was one of our major purposes in writing the text.

To our knowledge, no single text, or even a single discipline, has directly addressed all of the specific components of performance relevant to the needs of different clientele such as athletes or workers. It can be challenging to sort through various types of publications in many different disciplines to find information relevant to the assessment of clients' function. This text compiles information from many disciplines with the goal of covering all the parameters of functional performance and ways of testing for it. Thus the contents should be relevant to readers in many different areas.

Several forms of functional testing are available to the clinician. In our opinion, these tests are underutilized for several reasons: Clinicians are unaware of these tests, do not know where to find them, do not know how to interpret or use the information gleaned from them, or do not realize the importance of performing them.

Our purpose therefore was to provide a resource for clinicians to refer to in deciding how to assess functional parameters that are relevant to their clients (who could be patients, students, athletes, workers, etc.). While we recognize that no text is all-inclusive, we feel that this text can serve as a stand-alone resource or at the least as a supplement to other publications.

The organization of the text is user friendly. We explain both the "why" and the "what" of functional testing. The text is divided into three main parts. Part I provides background on what function includes and how it can be appropriately tested. Chapter 1 covers essential concepts and terminology related to functional testing, including a number of statistical terms. Readers will find the statistical terms useful when consulting the tables in part III of the text for information on the reliability and validity of the tests. Chapter 2 deals with the fundamentals of test administration—how to select the most appropriate test, how to decide on the order of testing, and how to avoid common problems and mistakes with testing. Chapter 3 provides specific examples and guidelines for implementing tests to best fit the needs of one's clients, whether in relation to sport, occupation, recreation, or fitness, and the working environment or situation.

Most of the text (parts II and III) is devoted to the actual assessment of functional performance. Part II describes testing procedures for discrete physical parameters such as anthropometry, flexibility, functional movements, and strength and power. Part III describes functional tests for specific regions of the body. Each of these performance-oriented tests is explained step by step. The following information is given for each test:

1. The purpose of the test
2. The equipment needed to administer the test
3. The testing procedure
4. Interpretation of results
5. Normative values for the test (if available)
6. Reliability and validity for the test (if available)

The appendix includes six reproducible forms that can be used to record data from selected tests.

We hope that readers will use many of these performance-oriented tests during examination of their clients. It is extremely important for clinicians to use functional tests to gain the best possible understanding of clients' actual functional limitations. Because of the lack of scientific data showing high correlations between standard tests or measures and functional outcomes, clinicians cannot fully understand the functional level of clients without using functional performance tests.

Readers will find the text easy to use whether they are in the gym with their athletes, in the training room or physical therapy clinic, or on the job site with a client. They will be able to refer quickly to a test and in many cases will not need to prepare extensively or use specialized equipment. It would also be appropriate to use the text in a more stringent laboratory setting. We hope that use of the text in various settings will be helpful to clinicians in a number of disciplines and therefore stimulate interest in interdisciplinary collaboration and future research on physical performance testing. We believe that such collaboration would encourage additional research interests, which would in turn benefit all disciplines and, more importantly, our clients or patients.

It was our intention to make the text useful for both the novice and the advanced clinician. The organization of the text will make it easy for the novice to use. Most of the testing procedures explained in the first two chapters of part II require only static measurements and thus are relatively easy to perform. When appropriate, in the subsequent chapters in parts II and III, we introduce less advanced tests before presenting more advanced tests.

Advanced clinicians will also find the text useful because of the details on interpretation of data and statistical analysis that are presented for many of the tests. These data should help advanced clinicians formulate or refine their opinions on the most judicious use of tests for their clients.

We stress that clinicians should choose tests based on their appropriateness to the needs of the individual client. The principles for organizing tests and the examples of testing algorithms explained in chapter 3 should help readers use the tests in this text to their best advantage. Different situations call for different tests or testing batteries. We encourage clinicians to use their clinical judgment and critical thinking to appraise each situation and determine the most appropriate test(s) as modeled in chapter 3. This clinical judgment and critical thinking will enable the most optimal use of the text.

To help readers understand the tests more fully we have also created a DVD. The DVD features live-action demonstrations of 40 of the most advanced tests. Icons on those test pages indicate they are modeled on the DVD. In addition, the DVD may be loaded onto a computer to offer convenient print-on-demand access to the reproducible forms in the appendix.

This text, with its reader-friendly format and broad range of content, was intended to reach a large audience. Clinicians will be able to use the text to identify suitable functional tests to incorporate into assessments for clients with sport, exercise, and occupational injuries. Fitness professionals will use the material for client assessments and motivation. Physicians, physical therapists, athletic trainers, and chiropractors may use selected tests to augment their test batteries when assessing patient progress.

Acknowledgments

We would like to acknowledge the following people:

- First and foremost, our parents and family members. Both of our parents have instilled in us a strong work ethic and the need for passion in our work. Neither of us would be where we are today without you!

- Our dean, Peter Cohen, PhD, as well as Cam Wilson, PT, PhD, and the rest of the faculty and staff at Wichita State University Physical Therapy Department and College of Health Professions for allowing us room to grow and expand outside of the college.

- Wichita State University physical therapy students, both past and present, for going along with our experiments. We are constantly learning from you!

- The staff at Human Kinetics including Loarn Robertson for first believing in our project and seeing it through; Maggie Schwarzentraub for continually rewording our apparent lack of understanding of the English language; and Gregg Henness and Neil Bernstein for making us look better than we really are on video and photo.

- Our models—Emily Stockman, Ryan Ayres, Olivia Martinez, and Carlos Rodriguez—for also making us look better than we really are.

- Matt Sanders, CSCS, *D, and the staff at Player Development Solutions for the use of their facilities and equipment. We know it was an inconvenience for all of you. We greatly appreciate your help and understanding!

- Multiple peers too numerous to mention for their guidance, support, and necessary criticism that make us better.

part

I

Basics of Functional Testing

This part of the text covers essential concepts and terms associated with functional testing, the fundamentals of test administration, and the implementation of functional testing in everyday practice. In chapter 1 we discuss what functional testing is and explain a key concept known as evidence-based practice, along with other concepts and terms related to testing that are important for proper use of the tests described later in the text.

Chapter 2, on test administration, deals with the purpose of testing and the selection and sequencing of tests. In this chapter we also discuss the importance of proper preparation of the client, outlining the current evidence on such topics as warm-up as well as the pros and cons of the different types of stretching. The chapter concludes with a discussion of common problems with testing.

Understanding the discrete and regional physical parameters that make up a client's function is only one component of functional testing. It is even more important, we feel, for clinicians to understand how to implement various tests to determine success or failure in the client's overall function. Chapter 3, "Integration of Functional Testing Into Everyday Practice," presents examples of functional testing plans for different types of clients. These examples will help readers understand not only how to use tests, but also what tests to use and when to use them. The examples will also help readers see how best to use the material in the remainder of the text. We encourage readers to follow these examples when appropriate, but also to develop their own testing algorithms according to the specific needs of individual clients. As we stress throughout the book and as these examples illustrate, having a testing plan is paramount.

Essential Concepts and Terms

In this chapter we discuss what functional performance testing is and explain a concept important to functional performance testing known as evidence-based testing. As compared to objective measures of particular components of physical function such as physical strength, balance, proprioception, muscle flexibility, or range of motion, functional performance testing looks at physical function, including performance abilities or limitations, within the context of everyday life activities. Evidence-based practice dictates that in choosing and carrying out functional tests, clinicians carefully consider and integrate what is known about these tests with the client's needs and desires in relation to their function.

Functional performance testing is used to determine people's ability to participate at the desired level in sport, an occupation, or recreation; when individuals have been injured, it is used to deter-

mine their ability to return to the activities they need or wish to engage in. Scores on these tests should help to predict the level at which someone can perform or the level he or she can safely return to. Since successful performance frequently requires multiple skills, clinicians often use multiple tests that will mimic the activities relevant to the individual client.

Clinicians should adhere to the tenets of evidence-based practice in planning and carrying out functional testing. This means that they should be familiar with the scientific evidence about tests that has been reported in the literature. Being able to interpret this evidence in order to apply it to clients involves understanding a number of concepts, such as the reference standard, reliability, the kappa statistic, correlations, validity, sensitivity, and specificity. In the second part of this chapter we explain these and other terms that are important to the interpretation of what is known about various functional tests. As we stress throughout the book, knowing why a given test will be useful depends on understanding its relevance to the present physical function of the client and to what the client needs or wants to do.

What Is Functional Performance Testing?

"Functional" can be defined in various ways. Functional activities have been defined as "those activities identified by an individual as essential to support physical, social, and psychological well-being and to create a personal sense of meaningful living" (American Physical Therapy Association 2001). Austin (2007) defines function as "any movement at the level of the person that is task related, goal oriented, environmentally germane and involves the integration of multiple body systems and structures." "Testing" is defined as using a set of problems to assess abilities. Therefore, "functional performance testing" means using a set of tests to determine performance abilities or functional limitations. A functional limitation is the inability to perform a particular activity at a normal level (American Physical Therapy Association 2001). "Normal" should be based on such factors as a client's age, sex, body type, and occupation. Normative values are highly dependent on the tasks in the specific sport or occupation and their associated demands or requirements. Normative values are determined via collection of data on large numbers of subjects performing a task.

More specifically for this book, functional testing is defined as using a variety of physical skills and tests to determine (1) one's ability to participate at the desired level in sport, an occupation, or recreation or to return to participation in a safe and timely manner without functional limitations and (2) one's ability to move through up to three planes of movement as assessed via nontraditional testing that provides qualitative and quantitative information related to specialized motions involved in sport, occupation, and exercises. Functional performance is an aspect of everyday life, whether for the elite athlete, the industrial worker, the homemaker, or anyone else. The commonality among all groups is that some aspect of performance is needed for them to be successful in performing their respective skills or duties.

Successful performance often requires multiple skills, and sometimes these multiple skills are difficult to assess and quantify. For this reason clinicians may use multiple functional performance tests depending on the set of physical attributes they need to assess. We believe that these tests or series of tests can be used to determine functional performance outcomes. Scores on functional performance tests should help to predict the level of activity at which a client can perform or to which the client can safely return. For clients who are reimbursed by third-party payers, functional performance outcome scores may provide justification for reimbursement or for the need for further rehabilitation or care, or conversely, justification that further care is not needed.

In the case of clients who have been injured, immediately after injury there is normally, at minimum, a brief period of impairment. Impairment is a dysfunction or a significant structural abnormality in a specific body part or system (Verbrugge & Jette 1994). Of importance in our context, the abnormality can have consequences for physical functioning (Verbrugge & Jette 1994). It is during the immediate postinjury period that the more classical objective measures of physical function are used, such as range of motion, anthropometry, muscle strength, reflexes, and joint integrity. These measures are somewhat generic in that taken as a whole, they indicate the ability of the body or system to perform work. But, although these clinical examination measures are important, they do not appear to correlate directly with actual physical function. This is the reason for functional performance testing. The lack of clinical correlation also makes it imperative that functional testing as described in this book be performed to ensure that the individual can safely return to the given activity.

Compared to the classical objective measures of physical function, functional performance testing more closely mimics physical activities that people wish to return to. For example, the fact that a person has full shoulder range of motion does not mean that he or she can throw a softball far enough to return to full unrestricted play. If this same person has full range of motion, full strength, and full neuromuscular control and additionally is able to achieve an excellent score on the Underkoffler softball throw without adverse symptoms, we are much more confident about the prospect of a safe return to unrestricted overhead throwing activities. Even though some feel that observing the performance of an individual's usual activities is time-consuming (Verbrugge & Jette 1994), many functional tests closely approximate the activities that people need or wish to do.

Physical performance tests have varying amounts of complexity and impose different levels of physical demand. The clinician should ensure that testing is at the appropriate level and occurs at the appropriate time for the individual client. Functional performance testing should not be done in a random, haphazard manner. As with any objective measurement used to assess an injured client, one should carefully evaluate the appropriateness of each test before introducing it. For example, a client who is only six weeks post anterior cruciate ligament reconstruction is not an appropriate candidate for performance of the single-leg hop test. This test would be far too aggressive. It would be more appropriate and much safer to use the single-leg timed stance. Davies and Zillmer's (2000) functional testing algorithm (FTA) is an objective, systematic, functional-based testing progression according to which the client must pass one level of functional performance testing before proceeding to a higher-level test. The FTA appears to provide a safe and effective way to perform a variety of testing maneuvers. Chapter 3 describes the use of the FTA in detail.

Evidence-Based Practice: Understanding Statistical Terminology and How to Apply the Information

Evidence-based practice (EBP) has been defined as "the conscientious and judicious use of current best evidence in making decisions about the care of individual patients" (Sackett et al. 1996). Central to the concept of EBP is the integration of evidence into the diagnoses and management of patients. Tests used for diagnostic purposes cannot be deemed simply good or bad; rather the test may provide important information for certain clients under certain conditions, but not for others.

Understanding how to implement EBP requires understand of the associated terminology. In this section we discuss a number of terms that are central to EBP and their relevance within the context of functional performance testing.

The reference standard is the criterion that best defines the condition of interest (Jaeschke et al.

1997). The reference standard should have demonstrated validity that justifies its use as a criterion measurement (Task Force on Standards for Measurement in Physical Therapy 1991).

Reliability refers to the degree of consistency with which an instrument or rater measures a particular attribute (Domholdt 2000). Measurements can be affected by random error. In determining the reliability of a measurement, we are determining the proportion of the measurement that is a true representation and the proportion that is the result of measurement error (Rothstein & Echternach 1999).

There are two main forms of reliability: intrarater reliability and interrater reliability. Intrarater reliability is the reliability of a test or measurement based on the degree of similarity of results obtained by one rater during different performances of a given test. Interrater reliability is the reliability of a test or measurement based on the degree of similarity of results obtained from different researchers using the same equipment and method. A large number of functional tests have been found to be extremely reliable. In our discussion of functional tests in subsequent chapters of the book, we provide reliability information where available.

The kappa statistic (κ) is a chance-corrected index of agreement that overcomes the problem of chance agreement when used with nominal and ordinal data (Laslett & Williams 1994). In theory, κ can be negative if agreement is worse than it would be expected to be by chance. In clinical reliability studies, κ usually varies between 0.00 and 1.00 (Portney & Watkins 1993). The κ statistic does not differentiate among disagreements; it assumes that all disagreements are of equal significance (Portney & Watkins 1993). Probably the strongest limitation of κ is that it is an analysis of exact agreement; that is, it treats agreement as an all-or-none phenomenon with no room for "close" agreement (Portney & Watkins 1993).

The correlation coefficient is a measure that determines the degree to which the movements of two variables are associated. Correlation coefficients are very sensitive to sample size, and statistical power can be relatively high even with smaller samples. A correlation coefficient should always be interpreted in relation to the size of the sample from which it was obtained. With a sufficient increase in sample size, almost any observed correlation value will be statistically significant, even if it is so small as to be a meaningless indicator of association (Portney & Watkins 1993).

To quantitatively describe the strength and direction of the relationship between two variables, the Pearson product–moment correlation coefficient (r) is used. Correlation coefficients are limited in their use as indices of agreement because they are designed to assess only two raters or two ratings (Portney & Watkins 1993) and because they are measures of covariance rather than reflecting agreement (Huijbregts 2002).

The intraclass correlation coefficient (ICC) is a reliability coefficient calculated with variance estimates obtained through analysis of variance (ANOVA) (Huijbregts 2002). The ICC has an advantage over correlation coefficients in that it does not require the same number of raters per subject, and it can be used for two or more ratings or raters (Huijbregts 2002). Because the ICC is an average

based on variance across all raters, nonagreement may involve all raters, some raters, or only one rater. The ICC can be considered an average correlation across raters and therefore does not represent the reliability of any individual rater (Ebel 1951; Portney & Watkins 1993). Table 1.1 provides value descriptions for these various reliability measures.

Fortunately, many studies reported in the literature have compared functional tests to tests of other commonly investigated parameters. Researchers have mainly sought to determine whether a variable or score in a functional test is predictive of a variable or score in another test. For example, many researchers have correlated hop or jump tests to isokinetic measures of strength; that is, they have asked whether the time on the stork test correlates with a measure of strength such as an isokinetic

TABLE 1.1

Value Descriptions of Various Reliability Measures

Value	Description
KAPPA (K) BENCHMARK VALUES (DATA FROM PORTNEY & WATKINS 1993)	
40	Poor to fair agreement
40-60	Moderate agreement
60-80	Substantial agreement
>80	Excellent agreement
100	Perfect agreement
PEARSON COEFFICIENT (R) BENCHMARK VALUES FOR HEALTH SCIENCES (DATA FROM PORTNEY & WATKINS 1993)	
0.00-0.25	Little or no relationship
0.25-0.50	Fair degree of relationship
0.50-0.75	Moderate to good relationship
>0.75	Good to excellent relationship
INTRACLASS CORRELATION COEFFICIENT BENCHMARK VALUES (DATA FROM PORTNEY & WATKINS 1993)	
<0.75	Poor to moderate reliability
>0.75	Good reliability
>90	Reasonable reliability for clinical measures
INTRACLASS CORRELATION COEFFICIENT BENCHMARK VALUES (DATA FROM FLEISS 1986)	
0.4 to 0.75	Fair to good reliability
>0.75	Excellent reliability

measure of peak torque or peak power. Correlation values may be high or low. In the chapters that follow we list correlations reported in the literature. Part of the process of deciding which functional test to use should involve determining which test seems to correlate most closely to the function or physical component that one wishes to test.

When reviewing correlations for the functional tests discussed in subsequent chapters, readers should keep in mind that a higher number (close to 1.0) indicates a higher or greater correlation and that a lower score (closer to 0.0) indicates a lower correlation. It is important to remember, though, that in the case of many functional tests, there are no correlation studies. This does not mean that they are not good tests; instead, the lack of studies may simply be a matter of too many tests and too little time. Although significant research has been done in this area, there is room for much more in the future.

Validity is the degree to which a study or test appropriately measures what it was intended to measure (Domholdt 2000). A test must be reliable to be valid, but a test does not have to be valid to be reliable. In the context of functional performance testing, valid tests are tests that measure the abilities vital to the given sport, occupation, or aspect of activity of daily living.

Measurement validity concerns the extent to which an instrument measures what it is intended to measure (Portney & Watkins 2000). Regarding functional tests, we must consider whether a given functional test is capable of discriminating among individuals with and without certain physical traits. For example, if we test a client with knee ligament instability and another without instability, are we certain that the single-leg balance test (stork test) can discriminate between the two clients? For most functional tests, research on validity is limited.

Most of the functional tests described in the following chapters demonstrate face validity, which indicates that an instrument appears to test what it is supposed to test and uses a plausible method for doing so (Portney & Watkins 2000). With the vertical jump test, we measure the distance that a client is able to reach when jumping as high as possible. The result is the score obtained when the client jumps. This is the least rigorous form of measurement validity (Portney & Watkins 2000). Many problems can be associated with tests that have only face validity. For example, as happens commonly, the client may not fully understand the instructions or may not be giving full effort.

The terms we discuss next are important especially in the allied health fields but can also be relevant in functional performance testing. Sensitivity refers to the ability of a test to detect those persons who actually have a disorder (as indicated by the reference standard). Some functional performance tests can be used to detect abnormality or dysfunction in a system, such as abnormal lower limb symmetry or piriformis dysfunction. Sensitivity has also been referred to as the true positive rate (Sackett et al. 2000). A highly sensitive test has relatively few false negative results and therefore points to the value of a negative result (Sackett 1992). SnNout is a mnemonic for "high sensitivity, negative, rules out" (Sackett et al. 1992). If the sensitivity of a test is high, a negative result is useful for ruling out a disorder or condition. High sensitivity does not address the value of a positive test (Fritz & Wainner 2001).

Specificity refers to the ability of a test to correctly identify people who do not have a disorder or condition (as indicated by the reference standard). In other words, it is the proportion of people who do not have the disorder who test negative. SpPin is a mnemonic for "high specificity, positive, rules in" (Sackett et al. 1992). If the specificity of a test is high, a positive result is useful for ruling in a disorder or condition. A highly specific test has relatively few false positive results and therefore points to the value of a positive test (Sackett 1992). High specificity does not address the value of a negative test.

Unfortunately, relatively few tests possess both high sensitivity and specificity. Knowledge of the sensitivity and specificity of a test can help clinicians refine clinical decision making by allowing them to weigh the relative value of positive or negative results (Fritz & Wainner 2001). Choosing the most appropriate test(s) is a new theme in the EBP clinical decision-making model.

Likelihood ratios (LR) can be either positive or negative. A positive LR indicates a shift in odds favoring the condition when the test is positive; a negative LR indicates a shift in probability that favors the absence of a disorder (Fritz & Wainner 2001). Tests with a large positive LR generally have high specificity because both values point to the usefulness of a positive test (Fritz & Wainner 2001). Small negative LR values correspond to high sensitivity (Fritz & Wainner 2001). For Likelihood Ratio Interpretation, see table 1.2.

Positive and negative predictive values, as well as accuracy, have also been used in evidence-based medicine to determine the strength of the test or

TABLE 1.2

Likelihood Ratio Interpretation

(+) Likelihood ratio (+LR)	(−) Likelihood ratio (−LR)	Probability interpretation of LR
>10	<0.1	Generates large and often conclusive shifts in probability
5-10	0.1-0.2	Generates moderate shifts in probability
2-5	0.2-0.5	Generates small, but sometimes important, shifts in probability
1-2	0.5-1	Alters probability to a small, and rarely important, degree

method described in the literature. Predictive values give clinicians an estimation of the probability of a test's being either positive or negative.

■ Positive predictive value (PPV): given a positive test result, the probability that the individual has a particular condition

■ Negative predictive value (NPV): given a negative test result, the probability that the individual does not have a particular condition

Overall accuracy refers to the proportion of people who are correctly diagnosed. Increasing testing accuracy requires careful planning and consideration of detail. The following are ways in which clinicians can increase accuracy:

■ Properly preparing the clients to be tested by helping them understand the test procedure

■ Allowing clients to perform a standardized practice (especially any unusual aspects of the test) before doing the test

■ Ensuring that clients adhere to pretesting instructions

■ Ensuring that the testing protocol is followed exactly

■ Using testers experienced with the specific testing protocols

■ Using consistent and calibrated equipment during testing

■ Keeping testers' motivation consistent for side-to-side comparisons, as well as client-to-client comparisons

Using more than one test or measure, if not several, can improve determinations of probability. A recent trend is to use tests in clusters to more accurately predict a positive or negative result. Clinical prediction rules are tools used by clinicians to determine the likelihood that a client is presenting with a given disorder, based on a number of variables that have been shown to have predictive validity (McGinn et al. 2000). Multiple tests are used as a group rather than as single tests or isolated multiple tests. The use of clusters of tests is advocated to allow one to predict the possibility of a positive or negative result from clinical testing.

Test Administration

Testing of functional ability is an inexact science. How does one test a functional task? The most appropriate way would be to test ability in the particular task (specific sport, specific job task, etc.). The best way to see if someone can successfully perform a task is to assess the person doing that task. This requires careful observation by a trained clinician with vast knowledge of anatomy, kinesiology, movement mechanics, and so on. The difficulty then lies in how one objectively quantifies the performance. Is it imperative that the task be objectively measured? To determine a person's readiness to return to his or her sport, occupation, or activities, ultimately one needs to assess the athlete or worker progressively and the individual needs to pass a testing paradigm.

Test administration should take into account specific aspects such as safety and organization. Careful planning should precede test administration. Test administrators should attend to detail in the setup and administration of the testing, as well as carefully instructing the client on the necessities

of the testing procedure to avoid confusion and to ensure reliability. The remainder of this chapter deals with specific aspects of administration: purpose of testing, selection of tests, sequence of tests, client preparation, and common problems with testing.

Purpose of Testing

Performance testing should serve one or more specific purposes. In the rehabilitation setting, one can follow a specifically designed outline for performance testing to demonstrate progress, determine dysfunction or weak links in performance, and determine a progression in the treatment plan. In the performance enhancement setting, testing can serve similar functions. This type of environment is also more conducive than the rehabilitation setting to large group testing. With large group testing it is advantageous to have specific testing times, environmental conditions, and so on. It is also a prerequisite to have an organized testing format in order to avoid lengthy testing procedures. In rehabilitation settings, testing can be implemented at various time frames in the program. This type of setting is more adaptable to changes in testing format and protocol in accordance with the specific needs of the client.

The assessments implemented should be necessary. Testing solely for testing purposes would seem counterintuitive. The test or assessment or the battery of tests should have purpose. Testing for purposes of assessing the effectiveness of a program should incorporate fundamental skills that will be necessary for the client. These skills can include fundamental skills (e.g., fundamental movement) through to advanced skills (e.g., jumping, agility). Other specific purposes of testing or assessment can include developing a specific exercise prescription, establishing training goals, increasing the client's motivation, evaluating a particular client's skills, developing normative value data, obtaining information for future research purposes (e.g., trying to determine specific injury risk factors or specific skill sets for success in specific sports), evaluating the rehabilitation or training program of individuals, determining whether a rehabilitation or training program is accomplishing what it was designed to accomplish, and determining if an individual is ready to return to work or the athletic field.

Selection of Tests

The criteria for selection of specific tests are variable. The clinician should determine specific criteria for each client. These criteria include safety, relevance, specificity, validity, accuracy, and practicality.

Safety

Safety entails ensuring that the client is able to undergo testing (see American College of Sports Medicine's *Guidelines for Exercise Testing and Prescription* 2006 for additional information). The specific level of testing should be appropriate for the particular rehabilitation client. The level of difficulty of the testing procedure should progress gradually whether the client is in the rehabilitation or the performance enhancement environment. For safety purposes, clients should perform a graded warm-up for all maximal-effort testing, warming up at 25%, 50%, 75%, and then 100% effort prior to the recording of their performance.

Relevance, Specificity, Validity, and Accuracy

Testing should be relevant to the client's particular sport, occupation, or activity. Just as with developing a specific training program, the development of a testing battery should take into account the principle of specificity. This includes energy-source specificity, muscle action specificity, muscle group specificity, and velocity specificity. Metabolism is highly specific to the intensity and duration of the sporting event, to the extent that excessive development of one type of fitness may have a profoundly detrimental effect on another type of fitness (Reiman 2006). Aerobic and anaerobic abilities are tested in completely different ways. Application of this concept should reflect the needs of clients in testing and their subsequent return to "full function" (see table 2.1). Harman (1994) discusses key points with regard to specificity of training that apply to testing as well:

▪ Training is the most effective when resistance exercises are similar to the sport movement (or movements) in which improvement is sought (target activity).

▪ Select exercises similar to the target activity with regard to joint movements and the directions of those movements.

▪ Joint ranges of motion in the training exercises should be at least as large as those of the target activity.

Validity and accuracy of testing are factors to consider. If the clinician is trying to determine a specific parameter (e.g., speed, agility, trunk endurance), then it is important to select tests that actually measure what they claim to measure. Validity of specific tests is often a concern because validity is based on an established "gold standard" test of the given specific parameter, which is often difficult to achieve in the area of human performance. There are multiple studies on many human performance parameters, but research on specific validity is limited.

Practicality

Practicality includes important considerations such as the location and availability of both the client and the testing location, as well as the duration and cost of tests. These are all important variables that one should consider before implementing a testing

TABLE 2.1

Sport Analysis Characteristics to Consider With Injured Athletes in Returning to Competition

Factors to consider	Description
Specific movements performed in the sport	Specific muscles involved and in what manner
	Joint angles and range of movement in the sport
	Type of muscle contraction
	Open versus closed kinetic chain movements
	Load requirements of the sport
	Velocity and speed requirements of the sport
Metabolic systems primarily involved	Adenosine triphosphate–phosphocreatine (ATP-PC) source
	Lactic acid source
	Oxygen source
Injury history and prevention	Site of present injury
	Sites of previous injury, injury history
	Most common sites of injury in the sport
	Anatomical and biomechanical factors of each athlete

This table is from an article published in *Postoperative orthopedic sports medicine: The knee and shoulder,* M.P. Reiman, Training for strength, power, and endurance, pg. 91, copyright Elsevier 2006.

battery. Another variable to take into account is the availability of the equipment required for testing as listed in this book for each test. Although "functional" testing often requires less equipment than traditional testing, some functional tests do require the use of specific types of equipment for proper performance.

Sequence of Tests

The sequencing of tests can determine success or failure in testing. General guidelines include performing the least fatiguing tests first and testing highly skilled tasks (agility, explosive ability, etc.) before fatiguing tasks. Endurance exercise preceding strength training appears to significantly decrease strength scores (Leveritt & Abernethy 1999). Aerobic training at a variety of durations and intensities has been shown to compromise both isotonic and isokinetic muscle strength performance at both 30 min and 4 h postexercise (Abernethy 1993; Leveritt & Abernethy 1999), while using a recovery period of 8 h was shown not to compromise strength performance (Leveritt et al. 2000). Although endurance testing is typically not for extended periods of time, these studies demonstrate the necessity of using proper testing sequence.

Testing is much easier to control with respect to time and test sequence in a rehabilitation setting than in other settings. Testing for a large group of athletes or occupational clients is ideally set up in stations in accordance with the sequence suggestions just outlined. Multiple testers are needed depending on the number of stations, number of clients, required rest intervals, and so on. To ensure accuracy of testing, one also needs to take other variables into account, such as level of supervision and detail of the testing protocol or testing battery. Organization is paramount to reliability in large group testing. Some deviations in sequence may be necessary for large groups but should be minimized. If proper rest is provided (at least 5 min between stations in order to restore the phosphagen energy system) (Harris et al. 1976) and if endurance and shuttle runs (the most fatiguing tests) are performed at the end of the testing sequence, the results are still likely to be accurate (Hoffman 2006).

Client Preparation

Prior to implementation, the client should be given a detailed explanation of any tests to be performed. Client preparation also includes a proper warm-up and practice trials.

Warm-Up

The client should perform an adequate warm-up. Stretching has long been viewed as a necessary component of an adequate warm-up routine. Three major types of stretching have been employed. Studies have generally indicated that proprioceptive neuromuscular facilitation stretching improves range of motion more effectively than either slow static or ballistic stretching (Anderson & Burke 1991; Etnyre & Abraham 1986; Holt et al. 1970; Shrier & Gossal 2000; Wallin et al. 1985), but this finding has recently been debated (Thacker et al. 2004). There are obviously pros and cons to each method, and clinicians should consider these when determining which type of stretching to implement (see table 2.2). The fact that the viscous properties of muscle resist elongation more when the stretch is applied rapidly (Taylor et al. 1990) suggests that greater strain is placed on the muscle with this type of stretching. This may in turn lead to microscopic tearing of the muscle fibers and connective tissue. As a consequence, nonelastic scar tissue would replace elastic muscle tissue and therefore long-term flexibility would decrease. Proprioceptive neuromuscular facilitation stretching typically requires the use of a partner, and the client does not have as much control over the amplitude and duration of the stretch as in static stretching.

TABLE 2.2

Comparison of Stretching Techniques

Factor	Ballistic	Slow static	PNF
Risk of injury	High	Low	Medium
Degree of pain	Medium	Low	High
Resistance to stretch	High	Low	Medium
Practicality (time and assistance needed)	Good	Excellent	Poor
Efficiency (energy consumption)	Poor	Excellent	Poor
Effective for increasing ROM	Good	Good	Excellent

PNF = proprioceptive neuromuscular facilitation; ROM = range of motion.

Reprinted, by permission, from V.H. Heyward, 2006, *Advanced fitness and exercise prescription*, 5th ed. (Champaign, IL: Human Kinetics), 266.

Some evidence suggests that stretching prior to an activity may be detrimental to performance (Knudson et al. 2000). A strength deficit may be seen for up to 1 h after intense, prolonged stretching (Fowles et al. 2000). It is thought that static stretching dampens the muscle response and does not allow the muscle to respond as quickly (Gray et al. 2002; Kubo et al. 2001; Church et al. 2001). To reach peak tension during contraction, stretched muscles need time to take up the slack in the musculotendinous unit that is produced by stretching. Stretching immediately before performance may impair strength and performance, especially when the muscle is not allowed enough time to take up the slack (Bracko 2002). There is also evidence that static stretching of the calf and thigh muscles reduces jumping performance and that a warm-up run or practice jumping improves concentric vertical jump performance height 3.4% ($p < 0.05$) (Young & Behm 2003).

There is also a lack of scientific evidence supporting the long-held belief that stretching prior to physical activity prevents the occurrence of injury (Pope et al. 2000; Thacker et al. 2004; Weldon & Hill 2003). Theories based on clinical observations and research data have been proposed to explain why stretching does not reduce the risk of injury (Heyward 2006):

- The ability of muscles to absorb energy is not related to flexibility. No scientific evidence supports the idea that more flexible muscles and connective tissues have a greater ability to absorb energy and are thus less likely to sustain injury (Shrier 1999).
- Even mild stretching can cause damage at the cellular level (Shrier 2000).
- The analgesic effect of stretching increases pain tolerance (Shrier & Gossal 2000).

Evidence would then seem to suggest that if stretching is used as part of a warm-up, it is necessary to avoid potential damage or fatigue that can be caused by overstretching (Weldon & Hill 2003). Complicating the issue is the fact that due to the analgesic effect of stretching, it may be difficult to perceive exactly where the end point of stretching (and therefore potential damage) occurs.

Dynamic movement types of stretching or activities (vs. no warm-up, static stretching, or dynamic stretching) have been advocated for preparation prior to testing or activity in order to prepare the body to avoid the dampening effect and to more quickly respond to the imposed demands with decreased risk of injury (Gray et al. 2002; Kubo et al. 2001; Church et al. 2001; Fletcher & Jones 2004; Young & Behm 2003; Koch et al. 2003). A goal of the dynamic warm-up should be to gradually assimilate the aspects of the test to be performed, or at least similar types of movements. Progression of difficulty allows clients to gradually increase their heart rate, body temperature, and so on, as well as their readiness to perform the required tasks.

It is difficult to generalize about stretching as an intervention. The role of flexibility in injury risk and prevention is also not clear. A client who is already extremely flexible may have little to gain from stretching, whereas clients with stiffer joints and muscles may well benefit. Further studies may help clarify these important and interesting questions.

It is relatively clear that excessive volume and load may result in fatigue if testing is to be implemented immediately afterward (Häkkinen 1993). A heavy resistance exercise stimulus designed to elicit an increase in posttetanus muscle twitch tension (postactivation potentiation) has been shown to do so under both isometric and dynamic test conditions. This increased muscle activation has typically been observed between 5 and 20 min following the stimulus (Gullich & Schmidtbleicher 1996; Houston et al. 1985; Smith et al. 2001). Postactivation potentiation following isometric and dynamic stimuli has been found to increase rate of force development, jumping height, and sprint cycle performance (Gullich & Schmidtbleicher 1996; Smith et al. 2001; Young et al. 1998), although the effect appears to depend on training status (Chiu et al. 2003). In athletically trained clients, the heavy resistance exercise stimulus enhanced power performance for 5 to 18.5 min, whereas recreationally trained individuals appeared to exhibit fatigue in the 5 min following the stimulus (Chiu et al. 2003). Whatever warm-up the clinician deems appropriate should be consistently implemented, not only for all clients but also for each client at each successive testing session. Eliminating any changes in the testing protocol will help ensure accuracy and consistency in testing.

Practice Trials

Since motor learning and strategy development are normally associated with practicing physical tests, a test may have to be repeated a number of times to yield reliable data (Hopkins et al. 2001; Jackson et al. 2001). The goal of administering test practice sessions is to enable clients to become proficient

enough that the test results are reliable and may then be used as a credible measure of their physical performance (Pandorf et al. 2003). It is also generally accepted that a learning effect occurs when people perform relatively new tasks and that initial strength gains are due to neural adaptations (Fleck & Kraemer 2004). Exposing clients to the testing procedures allows them to "learn" and become familiar with the movement. This helps to ensure maximal performance on the initial testing and hopefully also to minimize learning effects in later testing.

It is the clinician's responsibility to ensure that the client is ready for a specific level of testing. The clinician needs to consider several factors: any requirement for an informed consent form and release of liability form, the client's specific level of participation, the client's risk factors, the demands of the tests, risks and benefits of testing, and so on. Safety considerations are paramount with testing, especially with higher levels of testing and progressive testing within a session. Providing a safe testing environment, proper supervision, and proper warm-up and cool-down are just some of the multiple factors that the clinician needs to consider when implementing a testing program as part of a rehabilitation or performance enhancement program. If testing for fitness and performance enhancement, the clinician should consider some form of medical questionnaire and medical referral to a physician if this is deemed appropriate. Heyward (2006) and the American College of Sports Medicine (2000) provide information on the use of medical questionnaires and risk factor assessment prior to testing.

Common Problems With Testing

As with any type of testing, procedural problems may interfere with achieving the anticipated results.

■ Applying rules, guidelines, and procedures inconsistently. Use of consistent rules during testing is paramount. Inconsistency in the use of rules will lead to discrepancies in the results gathered during functional testing. Testing procedures and rules for testing are listed in the following chapters. These rules or guidelines have been developed to allow the use of a method that will ensure consistent testing between groups and over time. If testing is performed in an aberrant or inconsistent manner, the results may be useless. For example, suppose a basketball team is performing the double-leg horizontal jump for distance but that some of the players are allowed to use the upper extremities while others are required to hold their hands behind their back. It seems intuitive that those allowed to utilize their upper extremities during the jump will have better scores. Does this mean that they have better physical performance characteristics than the others? Probably not. Only consistent testing can show exactly who has better physical performance skills.

That being said, the scientific literature sometimes describes several ways of doing a particular test. For example, to perform the double-leg horizontal hop test, people can use their upper extremities, or keep their hands behind their back, or place their hands on their hips. Which way is best? There is no real consensus as to which way is best, and any answer is more than likely simply opinion. The easiest approach is for clinicians to decide which method they feel best addresses the component of physical ability they want to test and to use that method consistently.

■ Allowing unwanted compensation patterns. One of the more difficult aspects relating to clinical practice and incorporation of functional testing is unwanted compensation patterns. The human body has a remarkable way of altering motor patterns to fit the physical function available to the system being used. At times in physical rehabilitation of injured athletes, we actually want compensatory patterns. For example, the person undergoing conservative treatment following an anterior cruciate ligament reconstruction will probably be asked to selectively increase the activity of the hamstring muscles because of their agonistic relationship with the original ligament. This compensation pattern is highly sought after in this scenario. Another example is increasing compensatory strength of the rotator cuff and scapular muscles for the client who has an unstable shoulder. Again, this is a very functional and desired compensation pattern. But although some compensation patterns have clinical utility, faulty compensation patterns are not wanted. For example, clients who are performing the lunge test and do not lower themselves to the appropriate starting position are compensating adversely. If they are unable to perform this test without leaning forward excessively to recruit the gluteus maximus more, they are using an unwanted compensatory

pattern. Clients performing the single-leg horizontal hop for distance may use their arms when they have been asked to keep their hands on their hips; this is another example of an unwanted compensatory pattern.

■ If a client is giving 100% effort, the clinician can count on seeing some compensatory patterns. One way around this is to use clear and concise instructions in functional testing. The client may also need a visual description of the procedure if it is complicated. The complementary DVD provided with this book demonstrates proper technique for the functional testing procedures described in the text. As a final note on compensatory patterns, it is important to address them as soon as they are noticed. Unwanted and inappropriate motor patterns should not be allowed to continue during future sessions or testing. Reinforcing unwanted patterns only allows them to become more ingrained. If the patterns continue, there is a risk that the client will eventually feel that they are normal or natural. This is especially problematic in that once the patterns are ingrained into the motor system they become even harder to change. With continued use they may actually impede physical performance by placing the client at risk for further overuse injury.

■ Having clients perform numerous tests without adequate rest. Too often, a zeal to obtain as much information about a client as possible leads to the temptation to perform numerous functional tests in succession. With adequate rest between individual repetitions of tests, as well as between various functional tests, fatigue should not be a concern. For example, Manske and colleagues (2003) evaluated the reliability of functional tests following a closed kinetic chain isokinetic test and found excellent reliability for multiple tests including the double-leg vertical jump, single-leg vertical jump, two-legged horizontal jump, single-leg horizontal hop, and 6 m timed hop test. Physiologically, the timing of tests should be such that rest between them is adequate in accordance with the energy requirements for each one. Because the activities used by Manske and colleagues were anaerobic, relatively very short rest periods were needed (see table 2.3). Metabolically, these tasks require use of the phosphagen system; lactic acid will not accumulate in the muscle and thus a prolonged recovery time is not required. Most tests have some form of aerobic energy requirement, anaerobic energy requirement, or both. If a test appears to involve use of a dominant energy system, the clinician should determine this in advance to ensure appropriate test placement. For example, more rest would be needed following a longer-duration anaerobic-dominant test such as the Wingate bike test or the Margaria-Kalamen power test than following three repetitions of the double-leg horizontal jump for distance, which has a lower anaerobic requirement. Although all these tests assess anaerobic systems, the Wingate and Margaria-Kalamen Power Test last significantly longer and task the anaerobic capacity to a much greater extent than the double-leg horizontal jump, which appears to test peak anaerobic power.

■ Allowing too little or too much practice prior to testing. It is well known that motor learning effects occur during the performance of any type of motor activity, and they have been shown to occur during the assessment of functional testing maneuvers. We recommend a minimum of three practice repetitions to ensure adequate understanding of the testing procedure. This should also give clients adequate time to learn the technique and understand what is being asked of them. It is also clear that practice or repetitive utilization of a technique can enhance the results of a particular test. For example, it is probable that clients who use the Star Excursion Balance Test as part of their exercise program will score higher on this test than a person for whom the test is a totally novel activity. This would fit into the well-known concept of specificity of training in relation to improvement.

■ Using untrained testers. Functional testing is not a matter of simply watching someone jump, hop, or sprint. Critical analysis of functional testing requires great skill and knowledge. The clinician who performs functional testing must fully understand the varying stresses applied to the musculoskeletal and cardiovascular systems during performance of these activities. Considerations for functional testing require adequate knowledge regarding multiple aspects of human motion (Austin 2007), including the following:

- Common postures
- Movement patterns
- Amount of motion
- Speed of movement
- Nature and magnitude of force and resistance
- Dominant versus nondominant extremities
- Planes of motion
- Muscular activation patterns
- Joint-specific demands
- Symmetry or asymmetry of motion
- Unilateral versus bilateral demands

■ Selecting inappropriate tests. As clinicians determine which tests to use for a client, the most important question to ask is whether the test will measure what they want to measure. Is the test

TABLE 2.3

Major Characteristics of Primary Energy Systems

Primary source of energy	Duration	Work-to-rest ratios
ATP-PC (phosphagen)	First 20 to 30 seconds of exercise	• 1:12 to 1:20 for work at 90-100% of maximum power (0-10 s) • 1:3 to 1:5 for work at 75-90% of maximum power (15-30 s)
Phosphagens and anaerobic glycolysis	Major source of energy from 30 to 90 seconds of exercise	1:3 to 1:4
Anaerobic glycolysis and aerobic metabolism	From 90 to approximately 180 seconds of exercise	1:1 to 1:3
Oxygen (aerobic metabolism)	Predominates over other energy systems after the second minute of exercise	1:1 (longer duration, less intensity) to 1:3 (shorter duration, higher intensity)

valid? Is it reliable? What is the test's correlation to the client's sport or occupation? (Chapter 1 discusses these terms in detail.) Administering a test that doesn't measure what one wants to measure is a waste of time.

Functional testing is not without its problems; but with critical review, careful observation and assessment of the performance of each test, and application of consistent instruction during testing, a large majority of the problems can be resolved.

Integration of Functional Testing Into Everyday Practice

The assessment of any client can be as intensive as the clinician likes. Simply assessing strength with manual muscle testing and range of motion with a goniometer is an accurate method of indirectly assessing a client's actual physical function. But why not actually test the client's physical attributes through a variety of functional testing methods? Standard measurements of strength and range of motion assess total function, but again, only indirectly. Can we say that a client can play basketball, including performing physical skills such as running, jumping, and cutting, just because his or her quadriceps manual muscle test indicates full strength? Maybe not! If we as clinicians use tests that are more functional, we will have better insight

into the individual client's ability to engage in or return to a given activity. Without the use of functional tests, clinicians are simply guessing that the client can perform particular important activities. Functional testing should be used to the clinician's advantage to allow a better understanding of the client's actual physical capabilities.

Granted, as we have noted previously, "function" is a paradigm that is difficult to objectify. How do clinicians measure all of the physical attributes required to play a specific sport, play a specific position in that sport, or perform a specific task required in an occupational setting? They can closely monitor a client performing these specific tasks, but how do they objectively assign a grade or level to a given function? We have chosen to use the term "functional" for tests that are nontraditional, tests that are not typically part of a standard assessment as obtained via manual muscle testing, goniometry, and so on.

The functional tests discussed in this text can be used in multiple ways. A clinician can simply use any one or a number of the tests for a particular condition. For instance, to test a baseball pitcher's shoulder power, the clinician can utilize the medicine ball toss to determine the distance thrown. The same client can perform this test serially or can perform a given set of functional tests following a strengthening program to show whether or not the power training is working appropriately. In this way, objective criteria that are more functionally relevant can be used to document actual function-related clinical outcomes. Testing in most instances should be done in such a way that each of the components of the various physical attributes will be assessed. Physical function, as we interpret it, incorporates many attributes that can all be tested through the use of assessments from some or all of the categories covered in this book. Such testing could include assessment of balance and proprioception, muscular strength, muscular endurance, muscular power, speed and agility, aerobic and anaerobic conditioning, and functional movement patterns (figure 3.1). Depending on the client, flexibility may or may not be a necessary assessment. Another client may not need testing from the balance and proprioception category.

Which of the various tests in each physical attribute category to include is determined by the clinician in charge of the client's care. For some clients, only one or two attributes may require testing; for others, an entire battery of testing is warranted. As discussed in chapter 2, the testing sequence should progress from simple to more complex procedures and from less fatiguing to more fatiguing procedures—that is, from less demanding tests to more difficult, complex tests—so that the client does not become physically drained. For example, it would be appropriate to perform the double-leg jump test prior to the single-leg hop test, since landing and jumping with both legs is less stressful to either leg than jumping and landing with a single leg. As another example, the single-leg squat test should be done prior to the Lower Extremity Functional Test (LEFT), a very fatiguing test that in most instances should occur last in a sequence of tests.

Another excellent way to use these tests involves what is known as a functional testing algorithm (FTA) (Davies & Zillmer 2000). An FTA is a systematic, objective procedure that uses quantitative and qualitative criteria to allow a client to progress from one level of performance to the next-higher level. This appears to be especially important for an injured client who is in the process of physical rehabilitation. Using an FTA is also an excellent way to safely increase complexity while one is attempting to progress a noninjured client by enhancing physical function. With an FTA, clients are asked to perform a relatively simple test and prove that they have the appropriate strength, endurance, or other parameter addressed in that test before being asked to perform a more complex, higher-level test maneuver. Although Davies and Zillmer (2000) have described their own FTA, anyone can develop a continuum of functional tests that will benefit and guide an individual client through a rehabilitation program progression, or simply assess the various physical attributes needed by healthy clients in a sport, occupation, or fitness context.

We encourage readers to use the various testing procedures systematically in order to formulate their own FTAs in accordance with each client's needs. It would be impossible to describe every way in which an FTA can be developed; developing an FTA is a highly individualized means of organizing functional tests. The tests used in an FTA are based on the clinician's experience, knowledge base, skill, and preferred practice patterns. Recognize that not all clients will need all aspects of functional testing as described in this book. Many clients may only need several forms of functional tests to determine

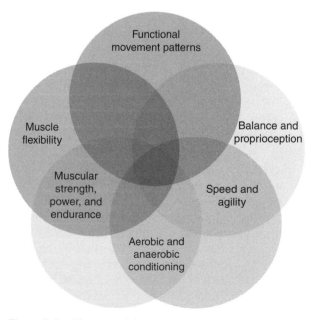

Figure 3.1 The essential components of functional testing.

their abilities. Also, different clients will require different aspects of functional testing, possibly even different aspects at different points in their rehabilitation program or training cycle.

In the following sections we provide examples of the use of functional tests in various situations. We use a case study to illustrate testing for a client recovering from a lower extremity injury, discuss preseason testing for a large group of athletes, and present a case study of a client in an occupational setting. We also outline some important considerations for preemployment screening and discuss testing in a clinical research or laboratory setting.

Client With Anterior Cruciate Ligament Repair

Natalie was a 17-year-old high school basketball player who was being seen following an anterior cruciate ligament reconstruction using a semitendinosus hamstring graft source. She had been in rehabilitation for 24 weeks following her reconstruction. At 12 weeks she had been given an isokinetic test for muscle strength and had been found to have 75% quadriceps strength and 85% hamstring strength in the involved compared to the uninvolved extremity at velocities of 180°/s and 240°/s. She was allowed to begin a jogging progression at that time and instructed to begin hopping and jumping activities with a gradual progression to full return to sport and recreational activities. She asked her physician to allow her to return to competitive basketball, and her physician wanted to make sure that she did not have any type of functional limitation.

Because this client was a patient following surgical reconstruction, a thorough evaluation to determine her full functional level was required, beginning with a complete history and subjective examination. The general physical examination consisted of basic examination measurements including visual analog pain scale scores, range of motion measurements, manual muscle testing, and kinesthetic and proprioceptive testing. See "Basic Measurements That Precede a Functional Testing Algorithm" on the right.

Several tests can be used to determine other basic measurements with an anterior cruciate ligament patient; these include the KT-1000 (MedMetrics Corp., San Diego, CA), which measures the amount of anterior tibial excursion. If the KT-1000 demonstrates less than a 3 mm difference bilaterally, the patient can undergo a variety of functional tests. If clinically available, isokinetic testing can be done to measure quadriceps and hamstring strength. As long as the involved extremity strength is within 25% of the uninvolved, functional testing can commence. Natalie was given the same test that she had performed at 12 weeks and showed 92% quadriceps strength and 89% hamstring strength compared to values for the contralateral side. Both the quadriceps and hamstrings exhibited appropriate levels of strength to allow a continuation of the FTA.

The flowchart in figure 3.2 shows Natalie's progression through the FTA. For a list of exercises to be performed by clients who need continued rehabilitation, see table 3.1.

To begin functional testing, clients can perform balance and proprioception using the single-leg stance (stork test), the Romberg test, or both. These are simple tests that determine static stability. A test that seems to assess more dynamic function is the Star Excursion Balance Test described in chapter 7

Basic Measurements That Precede a Functional Testing Algorithm

History and subjective examination
Objective examination

- Observation and posture
- Vital signs
- Gait evaluation
- Leg length measurements
- Referral and related joints
- Palpation
- Neurological examination
  - Sensation
  - Reflexes
- Balance, proprioception, kinesthesia
- Manual muscle testing
- Active range of motion
- Passive range of motion
- Joint integrity testing
- Flexibility tests
- Special tests
- Medical tests

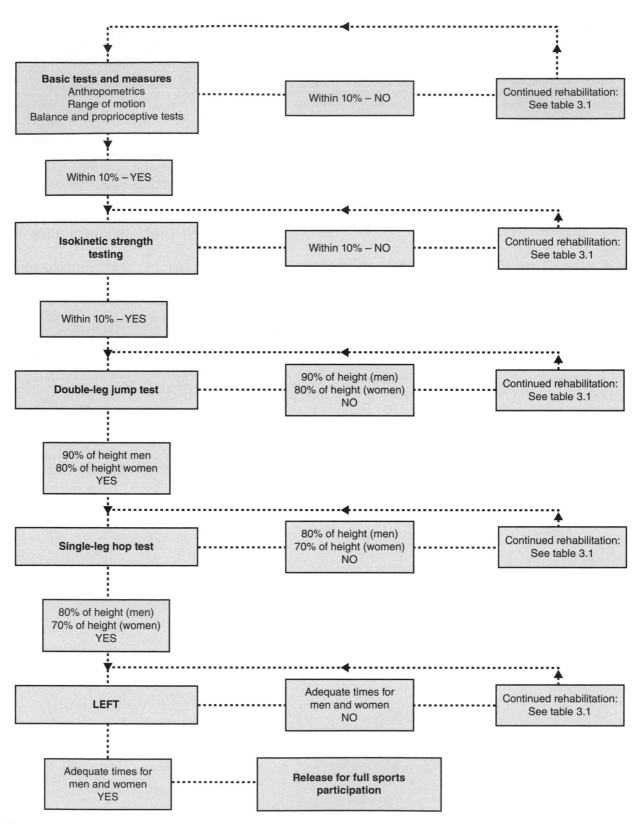

Figure 3.2 Functional testing algorithm following lower extremity injury.

TABLE 3.1

Exercises for Enhancement of Lower Extremity Testing Components

Balance	Strength	Speed and agility	Power
Single-leg stance	Leg press	Ladder drill	Plyometrics
Rocker board	Squat	Dot drill	Shuttle
BAPS board	Lunge	5-10-5 drill	Depth jump
DynaDisc	Knee extension	Carioca	Box jump
Sports beam	Hamstring curl	Cone drill	Olympic lifts
Mini-trampoline perturbation	Multiple hip exercises		

(Gribble and Hertel 2003; Plisky et al. 2006). This clinical functional test is purported to detect functional performance deficits associated with lower extremity pathology in either healthy or impaired clients. If clients using the involved leg can reach within 10% to 15% of the distance in all directions that they can reach using the uninvolved leg, they can progress to the next functional test. If the score for the involved leg is not within 10% to 15% of the uninvolved score, they should return for rehabilitation that emphasizes balance and stability drills. Single-leg balance and proprioceptive and perturbation drills can be used to enhance balance abilities. Proprioceptive-dominant exercises include use of the balance board, sport beam, DynaDisc, mini-trampoline, and the Biomechanical Ankle Platform System (BAPS).

The next functional test in the FTA is the double-leg jump for distance. This test can be done with or without the use of the upper extremities, although their use will probably help increase the distance jumped. Consistent testing is the key to reproducibility. The double-leg jump for distance is not only a test of the distance jumped; the performance also should be assessed subjectively for the quality of movement. Do clients land equally bilaterally, or do they compensate with the uninvolved extremity? Do they land awkwardly or with a loss of balance? The grading of the test can become somewhat subjective. Clients who can jump an appropriate distance but are unable to land correctly and who consistently fall forward, reaching for the floor with the arms to stabilize themselves as they fall, are probably not yet ready for the upcoming more stressful testing procedures. These clients may be better off continuing to work on small jumping and hopping drills, progressing from submaximal

to maximal jumps as they are able to tolerate. If clients can adequately stick the landing, the distance they are able to jump is standardized to their height. Owing to differing heights, leg lengths, and jumping abilities, absolute numbers used to record jumping distance can be misleading. Because of these limitations of absolute numbers, the jumping data are normalized to the client's height. Men are expected to jump 100% of their height while women are expected to jump 90% of their height. Clients unable to jump the required amount should continue basic rehabilitation with emphasis on double-leg exercises. Exercises to enhance ability to perform the double-leg jump include leg press, squatting, and lunging exercises.

If clients can adequately perform the double-leg jump and are able to jump a distance that is at least 90% of their height (men) or 80% to 90% of their height (women), they are allowed to attempt the single-leg hop test. This dramatically increases the difficulty factor, especially following lower extremity injury, because now the task is to land eccentrically on a single leg. As mentioned earlier, in the double-leg jump test, the client is asked to land on both legs and can compensate by putting more weight on the uninvolved side. With the single-leg hop, this is no longer possible. After leaving the ground by hopping off one leg, the client has to land on the single leg. As with the double-leg jumping test, this test can be done with or without the arms. Again as with the double-leg jump, performance on this test should not only be measured but should also be assessed subjectively for quality. An appropriate distance for the single-leg hop test is 80% to 90% of the client's height (men) and 70% to 80% of height (women). The bilateral comparison for the uninvolved extremity is 85% to 90%.

Normal scores for horizontal single-leg hop distance are from 143 to 203 cm in men and from 137 to 163 cm in women (van der Harst 2007; Ageberg et al. 2001). This large range is more than likely the result of varying procedures used in testing. Ageberg and colleagues (2001) allowed their subjects to use the upper extremities freely during the test, while van der Harst et al. (2007) restricted upper extremity movement. Ashby and Heegaard (2002) showed that the distance for the standing long jump without the use of arm movement is 21.2% less than with the use of free arm movements. Clients should be able to hop on the involved leg within 15% of the distance that they can hop on the uninvolved leg. Eighty-five percent limb symmetry has been described as normal (Noyes et al. 1991; Barber et al. 1990). Recent research has determined that there are no significant differences in single-leg hop distance between the dominant and nondominant legs of healthy athletes (van der Harst et al. 2007). If clients are unable to hop the height-normalized distance, they are asked to continue rehabilitation with concentration on single-leg exercises. Single-leg strength can be enhanced by the performance of single-leg squats, single-leg presses, and lunging exercises. Functional landing techniques are improved through the use of single-leg step and stick exercises and by double-leg takeoff to a single-leg landing and single-leg hops. These exercises require an eccentric load that is applied to the single leg. Eccentric loading is essential to the landing extremity in either of the single-leg hop tests.

The pinnacle of the FTA is the Lower Extremity Functional Test (LEFT) (Davies & Zillmer 2000), described in chapter 13. The LEFT is the most difficult test in the FTA because it incorporates many higher-level functional movement patterns including sprinting, retro-sprinting, side shuffles, cariocas, figure 8s, 45° angle cuts, 90° angle cuts, and crossover steps. As with most of the tests already described, the LEFT should be objectively and subjectively assessed. Hence a client with a fast time who does not perform the test properly would not pass. Because the LEFT is also an anaerobic test, a client with cardiorespiratory deficiency will have trouble finishing. It should also be stressed that during rehabilitation of injured clients, most likely only competitive or highly competitive recreational athletes would need to be evaluated using the LEFT.

A score on the LEFT of 100 s is average for men, <90 s is excellent, and >125 is poor. For women, a score of 135 s is average, <120 s is excellent, and >150 s is poor (Davies & Zillmer 2000). Clients

who perform poorly on the LEFT should return to aerobic conditioning and continue a maintenance program of strengthening exercises.

Baseball Team

Collegiate strength and conditioning coaches or athletic trainers may need to assess the physical abilities and attributes of their sporting teams. Functional tests can be conducted for the entire team or for individual groups of players depending on positions. For example, pitchers have different physical attributes than position players such as second base and shortstop. A variety of functional tests can be used to ensure that the testing involves the physical attributes germane to each position. See table 3.2 for a sample list of tests.

To begin, a test of balance and proprioception may be needed to determine if proprioceptive deficits exist. One may use the Star Excursion Balance Test or LEFT to test balance and proprioception (Hertel et al. 2000; Hertel et al. 2006; Kinzey & Armstrong 1998; Manske & Anderson 2004; Olmsted et al. 2002). Distance reached by the dominant and nondominant extremities should be within 10%. If deficits exist, it may be necessary to add neuromuscular balance and proprioceptive drills to the training. This can be done in multiple ways. One easy way to enhance proprioceptive and balance abilities is to complicate basic exercises with a pro-

TABLE 3.2

Sample Testing for College Baseball Team

Function	Tests	Interventions
Balance and proprioception	Star Excursion Lower extremity functional reach test	Lunge or squat exercises on disk or balance or tilt board
Strength	1RM squat	Leg press Lunge
Power	Vertical jump	Plyometric jump and hop training
Quickness, speed, and agility	T-test	Sprinting, carioca, change-of-direction drills

prioceptive component. For example, an athlete can do lunges or squat exercises on a disk or balance or tilt board. The athlete gains strength but also incurs a challenge to the proprioceptive system.

Strength can be assessed with the 1-repetition maximum (RM) squat as described in chapter 9, "Strength and Power." Normative values exist for normal healthy adults and specific athletic populations. If the athlete exhibits an overall decrease in general strength as assessed by the 1RM squat exercises, low to moderate repetitions and high loads should be the major source of training stimulation. Additional training could include leg presses and lunge exercises, although according to the principle of specificity of training, probably the best exercise to increase squat 1RM is the squat exercise itself. Specificity of actual sport functions would dictate that training should closely resemble the activities of the given sport, and more specifically the activities of the individual's position in the sport. Training for a particular sport should consider such variables as energy-source specificity, muscle action specificity, muscle group specificity, and velocity specificity (Reiman 2006). Different sports and different positions within a sport have specific demands for energy, muscle action and function, and velocity of muscle contraction.

Power can be determined with multiple tests, but one of the quickest and easiest is the vertical jump test. The athlete jumps next to a wall and the jump height is indicated with a marker or chalk, or the Vertec jumping apparatus (Sports Imports, Inc., Columbus, Ohio) can be used. Power training can be enhanced through the use of plyometric exercises including jump and hop training. This form of training is generally reserved for high-level athletes and if done improperly can result in injury or damage to musculoskeletal tissues.

Quickness, speed, and agility can be assessed with many tests depending on the specific parameter of speed, agility, or quickness that is relevant. A test that combines multiple attributes is the T-test, which is excellent for determining an athlete's or client's quickness, agility, and body control. Norms for the T-test are presented in chapter 10, "Speed, Agility, and Quickness Testing."

Client in an Occupational Setting

The progression of a client with a low back injury is outlined in the flowchart shown in figure 3.3. For a list of exercises to be performed by clients who need continued rehabilitation, see table 3.3.

Leonard, 38 years old, was diagnosed with L4-L5 lumbar radiculopathy. He was initially injured at work, and repetitive microtrauma was suspected because of his occupation as a warehouse worker at a local distributor. His job duty description included repetitive lifting of 20 lb (9 kg). He completed six weeks of rehabilitation and then rated his pain level as 0/10 at rest and 0-2/10 with work duties (which included a 15 lb [about 7 kg] weight restriction). At that point he had normal active range of motion of the trunk and hamstring flexibility within 10° bilaterally with the supine 90/90 test position (refer to chapter 5). Passive joint mobility of the lumbar spine improved to normal as compared to the restricted mobility noted at L3-L4 and L4-L5 with initial assessment. Although this manual assessment technique has shown poor to fair interexaminer reliability (van Trijffel et al. 2005), an assessment of overall and segmental spinal mobility was necessary. No dysfunction was noted with these tests and measures.

Leonard then was assessed with a Functional Movement Screen (FMS) (Cook et al. 1998). Refer to chapter 6 for specific testing parameters and interpretation of this tool. No pain was noted, and all scores were either a II or a III throughout the screen. Since Leonard consistently scored lower on the deep squat exercise of the screen, hip extension and abduction and trunk flexion movement analyses were performed. The hip abduction movement analysis showed altered muscle firing sequence (see chapter 6 for specifics). Additional motor control and strength and endurance exercises with emphasis on posterior gluteus medius were implemented (specifically clams, standing Thera-Band resistance walks, and side bridging exercises), with careful emphasis on monitoring for common compensations. Common compensations include trunk rotation with the side-lying clam exercise, excessive frontal plane motion with Thera-Band walks, and rotation of the body in the transverse plane with side bridging.

Once Leonard met the goal of normal muscle pattern firing with the same testing, he was given the Rockport Walk Test (see chapter 8 for specific testing procedures and other details). He walked the 1-mile distance in 11 min for a score of "Good." According to the established FTA for Leonard, this was an acceptable score, and he moved on to the trunk endurance test (see chapter 11 for details of these tests). It would be possible to improve cardiovascular endurance in a client such as Leonard with an aquatic program emphasizing endurance,

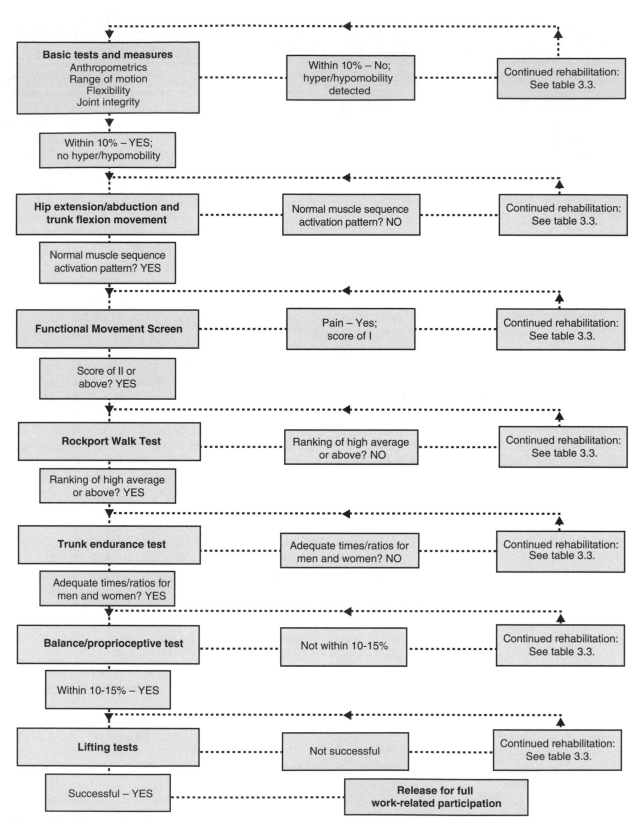

Figure 3.3 Functional testing algorithm following low back injury.

TABLE 3.3

Exercises for Enhancement of Low Back Testing Components

Functional movement and movement analyses	Strength	Cardiovascular endurance	Trunk endurance
Modified squatting activities	Squat	Aquatics	Isolated and global muscle retraining
Trunk flexion-extension activities (depending on the specific condition)	Lunge	Bicycle	Specific trunk stabilization exercises
Clam	Bridging	Walking	Integrated functional activities
Hip abduction in frontal plane only	Thera-Band	Elliptical machine	Lifting exercises for time (initially with low weight)
Isolated muscle contraction exercises with biofeedback			

a controlled walking program on a level surface or treadmill, and possibly use of an elliptical machine. The use of a bicycle is not necessarily prudent for a client with radicular complaints. In the initial trunk endurance assessment, Leonard scored in the normal range of the time frame for trunk flexion for normal individuals (McGill et al. 1999). He did not meet these previously established values for trunk extension endurance in normal individuals (people without pain or dysfunction). It is important to note, though, that these normative values were obtained with college-aged individuals. To our knowledge, no normative values have been established in patients with low back pain. A recent study (Flanagan & Kulig 2007) did reveal that 51.5% of participants after single-level microdiscectomy (from four to six weeks postsurgery) could not attain the fully extended position in a modified version of the trunk extension endurance test (Sorenson test). The authors concluded that the ability to attain the full trunk extension position was closely associated with fear–avoidance beliefs (Fritz et al. 2001), "suggesting this test may be too intense (either real or perceived) for many patients within 4 to 6 weeks following a single-level microdiscectomy."

Leonard returned to trunk endurance–specific activities including, but not limited to, specific local trunk endurance training with biofeedback and standing gluteal and latissimus muscle strengthening (modified squats, rowing-type activity for latissimus dorsi, etc.). These activities could be integrated into more functional training activities.

Trunk abdominal muscle contraction should be performed early in rehabilitation and progressed along with the functional activities used in the later rehabilitation stages.

After passing the trunk extension endurance testing (two weeks later), Leonard successfully completed the Star Excursion Balance Test as outlined in chapter 7 and discussed previously. He was able to reach in all directions with the involved extremity (where he had initially had radicular pain) within 10% to 15% of the distance possible with the uninvolved extremity.

The last assessment in the FTA established for Leonard was the repetitive box-lifting task (chapter 11). He successfully completed the initial attempt (four successful lifts). He was discharged from formal rehabilitation and referred back to his physician with the suggestion of release to full work-related participation. The testing results were faxed to his physician for perusal prior to his return visit.

This FTA was specifically established for Leonard because of his job title (warehouse worker) and his job duty description and requirements.

Specific activities that can be implemented to increase a client's ability to return to a repetitive lifting job duty could include proper squatting exercises (initially in one plane of movement; progression to all three planes with emphasis on hip mobility and trunk stability; progressive trunk stabilization exercises utilizing trunk abdominal muscle contraction; and actual lifting with emphasis on proper technique, and so on). Lifting tasks should initially emphasize low weight and higher repetitions to establish an endurance base. Specific lifting parameters (weight, distance, positions, etc.) can be integrated into the rehabilitation process on an individual basis when appropriate.

Client in Preemployment Screening

Assessment utilizing variations of the tests described throughout this text can be implemented in settings and situations other than postrehabilitation; an example is preemployment screening, which may be of particular importance for jobs with high physical demand levels.

Preemployment screening focuses on the physical skills required in the given job. Physical skills that are required of a firefighter, for example, may include running on level and unlevel surfaces; lifting and carrying heavy objects; pushing and pulling heavy objects; occasional jumping; balance; coordination; and quickness in multiple directions. Table 3.4, "Sample Preemployment Screening for a Firefighter," lists tests of functions that might typically be used for prospective firefighters.

As with any other battery of assessments, the clinician is advised to look critically at the particular requirements for successful performance in the individual's occupation and plan the assessment accordingly. The assessments and interventions should also be progressively advanced according to each individual's improvements and continued deficits.

TABLE 3.4

Sample Preemployment Screening for a Firefighter

Function	Tests	Interventions
Muscle length	Latissimus dorsi, pectoral muscles, and hip muscle lengths	Stretching for specified muscle length dysfunctions
Fundamental movement	Functional Movement Screen	Dynamic mobility drills as suggested in this chapter and other functional movement drills (e.g., hurdle step-over, lunge, balance drills)
Balance	Star Excursion Lower extremity functional reach test	Lunge or squat exercises on disk or balance or tilt board
Aerobic capacity	Multistage fitness test, 1.5-mile run test, or both	Bicycle Elliptical machine Treadmill walk/run Walk/run intervals
Strength	1RM squat	Leg press Lunge
Trunk endurance	Endurance tests for trunk extension, flexion, and bilateral side-bending	Isometric trunk endurance training progressing to dynamic trunk endurance tasks and trunk stabilization exercises
Lifting	Loaded forward reach or repetitive box-lifting task or both	Lunge Squat Lifting tasks Trunk endurance (as listed in preceding row) and strength interventions
Power	Vertical jump	Plyometric jump and hop training
Speed, agility, and quickness	T-test Illinois Agility Test	Sprinting in different directions and change-of-direction drills

The reader is encouraged to consult relevant sources (e.g., employee, employer, job description) in planning an appropriate testing battery for each client. Clients with preexisting dysfunction may also require specific modifications in tests. Sound clinical reasoning is important for the clinician planning a testing program.

Client in a Clinical Research or Laboratory Setting

The last example involves a client who might be seen in a fitness setting or in a clinical research or laboratory setting. In these settings, clients are often assessed for fitness levels. Some clients may want to know their fitness level as they begin an exercise program; others may want this information to use as a guide for progressing their workouts. Testing for these clients can potentially include tests from any or all the categories described in this book. See table 3.5 for a sample testing battery.

John is a competitive triathlete who is coming to the fitness center for tests to determine what his level of fitness is so that he can make changes to his present program in an attempt to improve in his sport. Testing for John begins with muscle length tests to determine mobility of commonly tightened muscle groups such as the quadriceps and hamstring muscles. Next, his aerobic capacity will be assessed via the 1.5-mile run test; this will probably be the physical parameter that he excels in since running the triathlon requires high levels of aerobic capacity. John would also like to know his anaerobic capacity, which will be measured via the Wingate test. Finally his lower extremity strength and power will be tested with the 1RM squat and with vertical and horizontal jumps and the horizontal hop test.

TABLE 3.5

Sample Testing in a Clinical Research or Laboratory Setting

Function	Tests
Muscle length	Quadriceps and hamstring muscle lengths
Aerobic capacity	1.5-mile run test
Anaerobic capacity	Wingate test
Strength	1RM squat
Power	Vertical jump, horizontal jump, hop, or a combination of these

If the fitness professional finds that John is weak in any of these characteristics, he or she can devise a program tailored to John's specific deficiencies. Such a program would be very helpful for someone who already trains and is highly conscientious about his fitness level but who would like to continue to see gains in specific parameters to help him meet his personal goal of improving triathlon times. Interventions will be selected after the testing identifies his present specific levels of fitness. He may be very proficient in some areas and need improvement in others.

As can be seen with these examples, functional testing can be used in a variety of situations. Functional testing can be implemented in an endless number of ways in the contexts of sport, occupation, and fitness. The tests described in this book are utilized to allow objective yet critical assessment of clients' physical performance and ultimate function. Throughout the book, our message to the responsible clinician is "Test—don't guess!"

part

II

Testing Procedures and Protocols for Discrete Physical Parameters

This section of the text deals with testing for specific physical parameters: anthropometric measures; flexibility; fundamental movements; balance; aerobic parameters; strength and power; and speed, agility, and quickness.

Each of these physical parameters is a component of overall functional ability, but each is required to a greater or lesser extent depending on the specific demands of the given functional task. The functional ability of most clients will consist of some, but not all, of these components, and to varying degrees.

In these chapters we outline the purpose of each test and describe in a step-by-step manner how it is to be used. We provide information on analysis and interpretation of data as well as on statistical analyses, when available, to help the clinician make sound clinical decisions on the appropriateness of a test in a particular situation. Some tests have been extensively investigated, others minimally or not at all.

This part of the text is organized from least to most complex, not only with respect to the physical parameters but also with respect to the tests within each chapter. We encourage clinicians to read the information about each test and to think critically in order to decide which tests are appropriate for each of their clients.

Anthropometric Assessment

Anthropometry is a series of systematized measuring techniques that express quantitatively the dimensions of the human body (Malina 1995). It has traditionally been used as a basic assessment of overall body size via height and weight. Anthropometry also has a long history in the physical education and sport science fields as a traditional assessment.

The anthropometric measurements utilized by the clinician are dependent on the goal of the assessment. Girth measurement and body mass index provide dimensions of stature that the clinician can utilize in various ways, including monitoring of weight loss and gain as well as body proportionality (waist-to-hip ratio). Measurement of trunk height could potentially be used along with the various trunk endurance tests described in chapter 11 to determine whether a correlation exists between greater trunk height and lower endurance scores (preliminary evidence obtained by the authors has not shown a significant relationship).

GIRTH ASSESSMENTS

▶ **Purpose:** Girth measurements are a simple means of measuring circumferences via specific landmarks on corresponding body parts.

▶ **Equipment:** Cloth tape measure, weight scale.

Procedure

1. Mark the sites to be measured.
2. With all of the measurements, use standardization of pull on tape measure—not too tight and not too loose.
3. Make sure that the tape is lying horizontal and is in contact with the skin whenever possible and appropriate.
4. Use the following procedures for specific sites:
   - Waist: Take measurement at the narrowest waist level, or if this is not apparent, at the midpoint between the lowest rib and the top of the iliac crest. Rudolf and colleagues (2007) suggest standardizing the waist circumference to 4 cm above the umbilicus.
   - Hip: Take measurement over minimal clothing and at the level of the greatest protrusion of the gluteal (buttock) muscles.
   - Chest: Take measurement at the level of the middle of the sternum, with the client's arms relaxed by the side, and at the end of a normal expiration.
   - Arm (relaxed): Have the arm relaxed and hanging by the side, and take the circumference at the level of the midpoint between the acromial (bony point of shoulder) and the olecranon (bony point of elbow) processes.
   - Arm (flexed): Have the arm raised to a horizontal position in the sagittal (forward) plane, with the elbow at a right angle. Ask the client to maximally contract the biceps muscle, and measure the largest circumference.
   - Thigh: Have the client stand with legs slightly parted. Take the circumference measure 2 cm below the gluteal line (buttock crease) with the tape horizontal.

Analysis and Interpretation of Data

High waist girths are associated with low back pain in both men and women (Albert et al. 2001).

Statistics

- Reliability: Factors affecting reliability could include different testers, tension on the tape (constant and even or not), and determination of anatomical sites (correct or not).
- Validity: These measurements are not a valid predictor of body fat. They are, however, a good measure of proportionality.

BODY MASS INDEX ASSESSMENT

▶ **Purpose:** The body mass index (BMI) is a simple measurement to help determine relationship of body height and mass.

▶ **Equipment:** Cloth tape measure, weight scale.

Procedure

1. Obtain client's body weight in kilograms (kg) and height in meters (m).
2. Calculate BMI using the following formula:
   BMI = Body weight (kg) / Height (m^2).

Analysis and Interpretation of Data

■ Normative values have been established by age and gender as a potential reference for use in interpretation of BMI data. See table 4.1 on page 36 and table 4.2 on page 37.

■ The relationship between BMI and fat as a percentage of body weight is approximately linear (Norgan 1994).

■ Although different cutoffs have been used, health risks for males and females begin when BMI exceeds ~25; if BMI is greater than 30, a person is judged to be obese (Foss & Keteyian 1998). Underweight has been generally accepted as a BMI less than 18.5, and a normal range is 18.5 to 29.9 (Foss & Keteyian 1998). It has also been generally accepted that as BMI increases, mortality rate from cardiovascular disease and diabetes increases (Smith & Haslam 2007).

■ Interestingly, not all information regarding a higher BMI is negative. A significant positive correlation was found between volleyball players' BMI and spike velocity (i.e., the heavier the player was in proportion to height, the more efficient he was in spike activity) in a study of 19 front-line male volleyball players in Belgium (Forthomme et al. 2005).

■ The results just mentioned could relate to the fact that a person with a large muscle mass will also have a large BMI. This is one of the problems with the use of BMI numbers in the absence of data about body fat percentages obtained with calipers or by other means. Body mass index does not differentiate between lean body mass and fat mass. Therefore, it is not appropriate to use BMI with an athletic population. For example, Shaquille O'Neal at 2.16 m and 148 kg has a BMI of almost 32 but is not obese. Other individuals with large muscle mass with demonstrated low percent body fats can have BMIs over 30 yet are obviously not obese.

■ The use of BMI during the transition into puberty and in adolescent males may have limitations due to the potential for significant gains in muscle mass and the constantly altering relationship between height and weight in these individuals (Malina 1995).

Statistics

Body mass index is moderately correlated ($r = 0.60$-0.82) with percent fat (Smalley et al. 1990).

WAIST-TO-HIP RATIO ASSESSMENT

▶ **Purpose:** This measure is often used to determine the coronary artery disease risk factor associated with obesity. The basis of the measure as a coronary disease risk factor is the assumption that fat stored around the waist poses a greater risk to health than fat stored elsewhere.

▶ **Equipment:** Tape measure and calculator.

Procedure (Bray & Gray 1988)

1. Obtain the client's waist and hip girths as already explained.

2. Perform the following calculation:
   Waist-to-hip ratio = Waist girth / Hip girth.

Analysis and Interpretation of Data

There are no known peer-reviewed published normative data for this assessment. There is a table of unpublished normative values (table 4.3 on p. 37) that should be interpreted with caution.

Statistics

There are no known published reliability data for this assessment.

TORSO HEIGHT ASSESSMENT

▶ **Purpose:** This is a simple method of determining the height of the client's torso. The assessment demonstrates potential benefit for comparison to other body parts, as well as for use in testing the relationship to trunk endurance (see section on analysis and interpretation).

▶ **Equipment:** Tape measure, athletic tape, string.

Procedure (Gross et al. 2000)

1. Have the client sit on edge of table, with his or her back as straight as possible.

2. Apply athletic tape over the spinous processes of the client's entire lumbar spine.

3. Wrap string around the index fingers of both your hands and palpate the surface of the bilateral iliac crest (the apex) with your index fingers.

4. A second clinician then approximates the string to the athletic tape running over the spinous processes and places a mark over this intersection with a marker.

5. Measure the distance from the apex of the client's head to the mark on the athletic tape. Record the measurement to the nearest 0.5 cm.

Analysis and Interpretation of Data

Trunk length analysis was not provided in the original reference (Gross et al. 2000). It is assumed that this measurement could have potential use in future correlation studies (e.g., trunk length and trunk endurance).

Statistics

	ICC [Intraclass correlation coefficient (2,1)]	Mean absolute difference
Torso height (cm) ($n = 28$)	0.88	0.8 ± 0.9

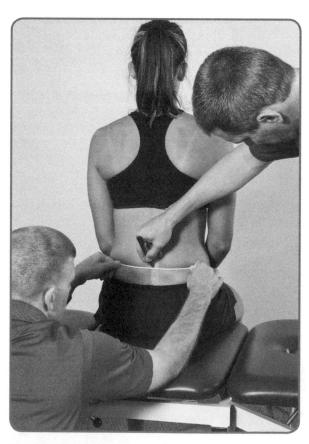

Marking intersection of tape and string.

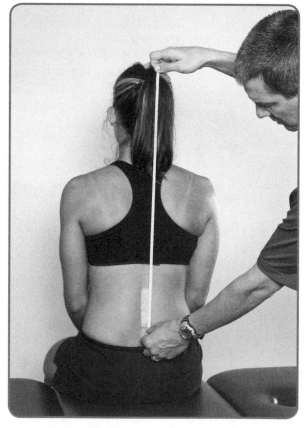

Measuring distance from apex of client's head to mark.

PELVIC GIRTH ASSESSMENT

▶ **Purpose:** To assess the girth of the pelvis.

▶ **Equipment:** Cloth athletic tape, flexible tape measure.

Procedure (Gross et al. 2000)

1. Place a strip of athletic tape vertically between the lateral aspect of the iliac crest and the greater trochanter on bilateral lower extremities.

2. Mark the most superolateral aspect of the iliac crest and the most lateral aspect of the greater trochanter.

3. Use the flexible tape measure to determine the girth of the pelvis at this superior lateral location. Pull the tape measure snugly over the individual's shorts, and record the measurement to the nearest 0.5 cm.

Analysis and Interpretation of Data

Girth measurements have been directly correlated with the ability to produce joint moment (Gross et al. 1990, 1989).

Statistics

	ICC [Intraclass correlation coefficient (2,1)]	Mean absolute difference
Pelvic girth (cm) *(n = 28)*	0.94	1.5 ± 1.4

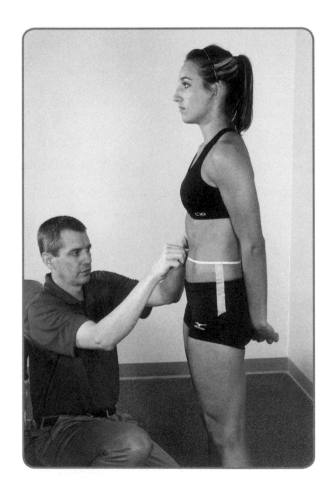

TABLE 4.1

Body Mass Index Chart

BMI	NORMAL (HEALTHY) WEIGHT						OVERWEIGHT					OBESE					
	19	20	21	22	23	24	25	26	27	28	29	30	31	32	33	34	35
Ht (in.)	BODY WEIGHT (POUNDS)																
58	91	96	100	105	110	115	119	124	129	134	138	143	148	153	158	162	167
59	94	99	104	109	114	119	124	128	133	138	143	148	153	158	163	168	173
60	97	102	107	112	118	123	128	133	138	143	148	153	158	163	168	174	179
61	100	106	111	116	122	127	132	137	143	148	153	158	164	169	174	180	185
62	104	109	115	120	126	131	136	142	147	153	158	164	169	175	180	186	191
63	107	113	118	124	130	135	141	146	152	158	163	169	175	180	186	191	197
64	110	116	122	128	134	140	145	151	157	163	169	174	180	186	192	197	204
65	114	120	126	132	138	144	150	156	162	168	184	180	186	192	198	204	210
66	118	124	130	136	142	148	155	161	167	173	179	186	192	198	204	210	216
67	121	127	134	140	146	153	159	166	172	178	185	191	198	204	211	217	223
68	125	131	138	144	151	158	164	171	177	184	190	197	203	210	216	223	230
69	128	135	142	149	155	162	169	176	182	189	196	203	209	216	223	230	236
70	132	139	148	153	160	167	174	181	186	195	202	209	216	222	229	236	243
71	136	143	150	157	165	172	179	186	193	200	208	215	222	229	236	243	250
72	140	147	154	162	169	177	184	191	199	206	213	221	228	235	242	250	258
73	144	151	159	166	174	182	189	197	204	212	219	227	235	242	250	257	265
74	148	155	164	171	179	186	194	202	210	218	225	233	241	249	256	264	272
75	152	160	168	176	184	192	200	208	216	224	232	240	248	256	264	272	279
76	156	164	172	180	189	197	205	213	221	230	238	246	254	263	271	279	287

Adapted from N. Payne et al., 2000, "Canadian musculoskeletal fitness norms," *Canadian Journal of Applied Physiology* 25:430-442. With permission from NRC Research Press.

TABLE 4.2

Canadian Body Mass Index Normative Data by Gender and Age (Mean ± SEM)

Age group (years)	n	BMI (kg/m^2)
FEMALES		
15-19	59	22.2 ± 0.4
20-29	83	23.7 ± 0.5
30-39	56	24.3 ± 0.6
40-49	47	24.1 ± 0.6
50-59	47	25.4 ± 0.7
60-69	20	27.3 ± 1.3
MALES		
15-19	54	23.1 ± 0.5
20-29	73	25.3 ± 0.4
30-39	44	26.3 ± 0.6
40-49	27	25.4 ± 0.5
50-59	36	26.4 ± 0.6
60-69	25	28.6 ± 0.8

Reprinted, by permission, from V.H. Heyward, 2006, *Advanced fitness and exercise prescription,* 5th ed. (Champaign, IL: Human Kinetics), 226.

TABLE 4.3

Waist-to-Hip Normative Values

	ACCEPTABLE VALUES		UNACCEPTABLE VALUES		
	Excellent	**Good**	**Average**	**High**	**Extreme**
Males	<0.85	0.85-0.90	0.90-0.95	0.95-1.00	>1.00
Females	<0.75	0.75-0.80	0.80-0.85	0.85-0.90	>0.90

5

Muscle Length Assessment

Flexibility can be defined as the ability to move joints fluidly through complete ranges of motion without injury (Heyward 2006). While flexibility is probably not regarded as a major parameter for assessment in human performance, it is integral to dynamic human movement and pain-free function. While the controversy regarding the use of static stretching as part of a warm-up routine for increased performance and injury prevention seems to lack merit (see chapter 2), there is some clinical evidence to suggest that prolonged stretching performed outside of the preexercise period can reduce incidence of injury (Hartig & Henderson 1999). Prolonged low-load stretching in animals has been shown to increase muscle length and hypertrophy (Goldspink et al. 1995; Lederman 1997; Yang et al. 1997) as well as permanently lengthening connective tissue (Sapega et al. 1981). Similar results have been achieved in human subjects with osteoarthritic hips (Leivseth et al. 1989) and joint contractures (Wessling et al. 1987). Evidence of similar effects in

healthy shortened muscles has yet to be obtained (Weldon & Hill 2003).

Optimal body alignment and proper agonistic–antagonistic muscle relationships have been advocated by many experts (Janda 1994; Sahrmann 2002; Kendall et al. 2005). Functionally shortened muscles have the capability of reciprocally inhibiting their antagonistic counterpart (Janda 1994; Sahrmann 2002; Kendall et al. 2005). This can lead to increased potential for future injury and less than optimal function. The terms "upper crossed syndrome" and "lower crossed syndrome" (Janda 1994) have been used to refer to such imbalances. Specific muscles have been shown to have a predisposition for tightness (pectoralis major and minor, upper trapezius, suboccipital muscles, hip adductors and flexors, erector spinae, etc.) and thereby inhibit their antagonistic muscle groups (lower trapezius, gluteal muscles, deep neck flexors, rectus abdominis, etc.). The clients display typical postures (Janda 1994) as a result of muscle imbalance between antagonistic muscle groups. The clinician should assess for these muscle imbalances by not only muscle strength testing, but also muscle length testing.

To our knowledge, information on normative values with respect to flexibility throughout the body is very limited. A dearth in the area of normative values for lower extremity muscle lengths in college-age populations has also been reported (Corkery et al. 2007). This brings up the question of what normal is. Normative values should be based on a client's age, sex, body type, occupation, and so on. Clients in sedentary occupations have lower flexibility demands than does a ballet dancer or a sprinter. Normative values should be representative of the specific clientele. This is most likely a primary reason for the paucity of normative value data in the current literature. In this chapter we outline information on muscle length testing that

appears in the literature and present reliability data when available.

We list the specific muscle lengths to be assessed in a systematic order: the upper extremity muscle measurements first, followed by the trunk and lower extremity muscle measurements.

The purpose of each measurement is to determine the extensibility, length, and excursion of the muscle. Comparison can be made to the muscle on the other side or to any normal values that have been established in the literature. Possible reasons for the scarcity of information on normal values are the difficulty in obtaining consensus on testing methods and the variability among clients in terms of age, sex, body type, sport or occupation, and so on.

For the origin, insertion, action, and innervation of the muscles covered in this chapter, the reader is referred to the tables at the end of the chapter (pp. 77-84).

Upper Extremity

LEVATOR SCAPULAE ASSESSMENT

▶ **Purpose:** To determine the extensibility, length, and excursion of the levator scapulae. For the origin, insertion, action, and innervation of this muscle, see table 5.1 on page 77.

▶ **Equipment:** Stable testing surface.

Procedure (Greenman 1996)

1. Position the client so that he or she is supine with arms initially at side.
2. Sit at the head of the table.
3. Stabilize the shoulder girdle with your lateral hand.
4. With your medial hand, support the occiput and introduce flexion, side-bending, and rotation away to the resistance barrier (position 1).
5. If you wish to use an alternative (position 2), place the client's arm above the head and exert caudad compression on the elbow with the same head position.
6. Assess the muscle length and compare to the opposite side.

Analysis and Interpretation of Data

▪ Asymmetry implicates restricted muscle length and excursion.

▪ There are no known normative data for this flexibility measure.

Statistics

There are no known reliability data for this flexibility measure.

Note

You can use contract-relax of shoulder elevation by the client against your resistance at the barrier to increase range of motion (ROM).

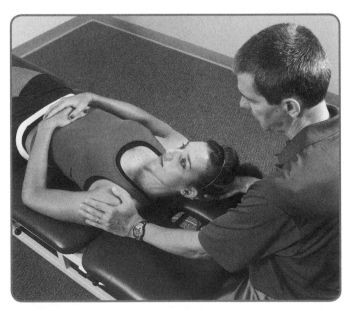

Position 1.

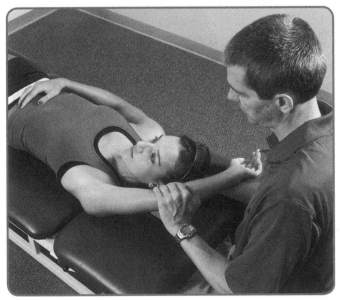

Position 2.

UPPER TRAPEZIUS ASSESSMENT—SITTING

▶ **Purpose:** To determine the extensibility, length, and excursion of the upper trapezius during sitting. For the origin, insertion, action, and innervation of this muscle, see table 5.1 on page 77.

▶ **Equipment:** Stable testing surface.

Procedure (Greenman 1996)

1. Position the client so that he or she is sitting in upright posture.
2. Stand just behind the side to be tested.
3. Passively laterally flex away and rotate the cervical spine toward the testing side (first position).
4. Passively elevate the shoulder girdle on the testing side (second position).

Analysis and Interpretation of Data

■ If additional ROM is achieved with shoulder girdle elevation, especially with cervical rotation, tightness is implicated.

■ Asymmetry from side to side also implicates upper trapezius tightness or restriction.

Statistics

There are no known reliability data for this flexibility measure.

Notes

■ This test can also be used to rule out cervical spine articular involvement.

■ Monitor trunk lateral flexion compensation when elevating the shoulder girdle.

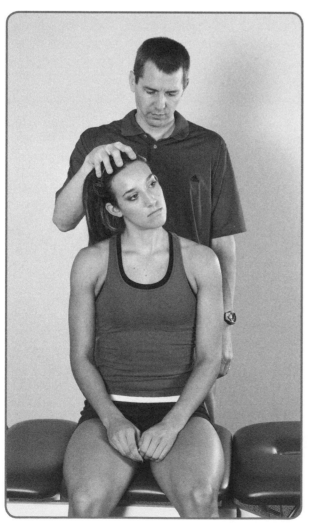

First position.

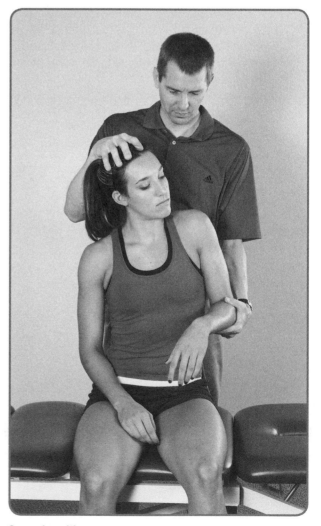

Second position.

UPPER TRAPEZIUS ASSESSMENT—SUPINE

▶ **Purpose:** To determine the extensibility, length, and excursion of the upper trapezius in the supine position. For the origin, insertion, action, and innervation of this muscle, see table 5.1 on page 77.

▶ **Equipment:** Stable testing surface.

Procedure (Greenman 1996)

1. Position the client so that he or she is supine with arms at side.
2. Sit at the head of the table.
3. With your lateral hand, stabilize the shoulder girdle to prevent elevation.
4. With your medial hand, support the occiput and introduce cervical lateral flexion away, with rotation ipsilateral to the side being tested.
5. Assess the muscle length and compare to the opposite side.

Analysis and Interpretation of Data

Asymmetry implicates restricted muscle length and excursion.

Statistics

There are no known reliability data for this flexibility measure.

Notes

■ You can alternate hand positions.

■ You can use a similar position to assess anterior and medial scalene muscles. Have the client's head off the end of the table and translate the head toward the floor (slight extension) with lateral flexion away and rotation toward the side being tested.

■ You can use contract-relax of shoulder elevation by the client against your resistance at the barrier to increase ROM.

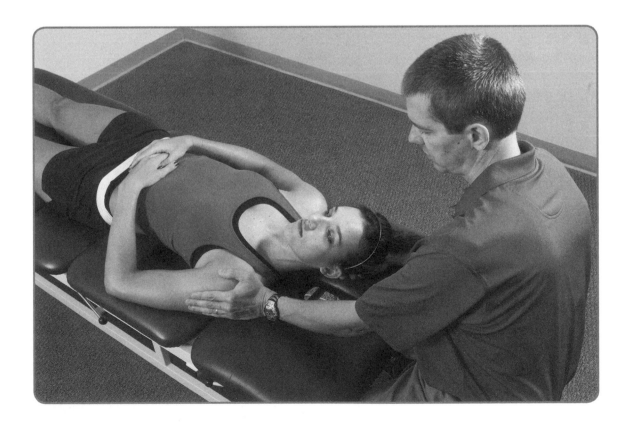

STERNOCLEIDOMASTOID ASSESSMENT

▶ **Purpose:** To determine the extensibility, length, and excursion of the sternocleidomastoid. For the origin, insertion, action, and innervation of this muscle, see table 5.1 on page 77.

▶ **Equipment:** Stable testing surface.

Procedure (Greenman 1996)

1. Position the client so that he or she is supine with arms at side.
2. Sit at the head of the table.
3. With your lateral hand, stabilize the shoulder girdle to prevent elevation.
4. With your medial hand, support the client's occiput and introduce cervical spine side-bending away, with rotation ipsilateral to the side being tested.
5. Assess the muscle length and compare to the opposite side.

Analysis and Interpretation of Data

Asymmetry implicates restricted muscle length and excursion.

Statistics

There are no known reliability data for this flexibility measure.

Notes

■ Emphasize client's chin tuck with cervical spine flexion and rotation in comparison to upper trapezius.
■ You can alternate hand positions.
■ You can use contract-relax of shoulder elevation by the client against your resistance at the barrier to increase ROM.

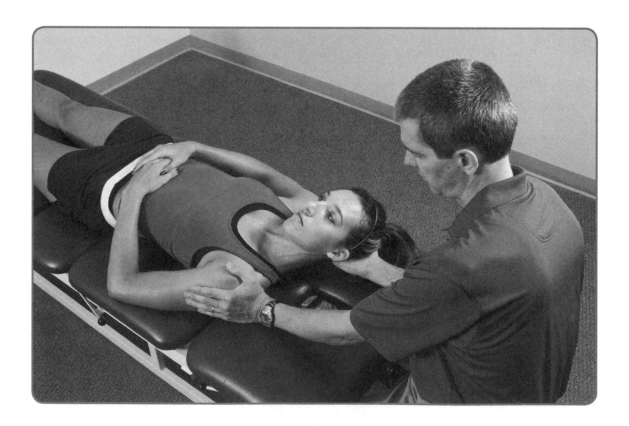

SUBOCCIPITAL MUSCLES ASSESSMENT

▶ **Purpose:** To determine the extensibility, length, and excursion of the suboccipital muscles. For the origin, insertion, action, and innervation of these muscles, see table 5.2 on page 77.

▶ **Equipment:** Stable testing surface.

Procedure (Flynn et al. 2000)

1. Position the client so that he or she is supine.
2. Either sit or stand at the head of the table and support the client's head with one hand on forehead and the other hand on occiput (position 1).
3. Flex the upper cervical spine (chin tuck) and apply gentle traction.
4. Rotate the head approximately 30° to focus the assessment on that side (position 2).
5. Assess the muscle length and compare to the opposite side.

Analysis and Interpretation of Data

Asymmetry implicates restricted muscle length and excursion.

Statistics

There are no known reliability data for this flexibility measure.

Notes

■ Providing adequate support of head at the occiput will allow the client to relax and provide for more reliable results.

■ This will also allow you to use your hand at the forehead to guide chin tuck (gently).

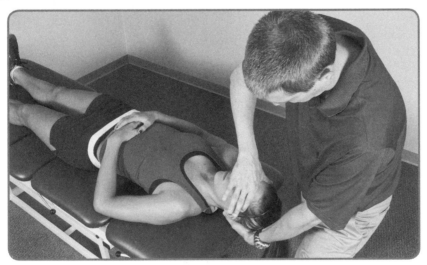

Position 1.

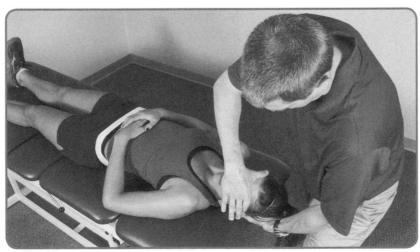

Position 2.

LATISSIMUS DORSI ASSESSMENT

▶ **Purpose:** To determine the extensibility, length, and excursion of the latissimus dorsi. For the origin, insertion, action, and innervation of these muscles, see table 5.4 on pages 78-79.

▶ **Equipment:** Stable testing surface, standard goniometer or standard tape measure.

Procedure (Reese & Bandy 2002)

1. Position the client so that he or she is supine with upper extremities at sides and elbows extended.
2. Ensure that the lumbar spine remains flat against the supporting surface.
3. Alongside the client, flex the client's shoulder through available ROM while maintaining the elbow in full extension and keeping the arm close to the head.
4. Ensure that the lumbar spine remains flat against the table surface.
5. If using the goniometer, place the axis at the shoulder glenohumeral joint, place the stable arm horizontally along the midaxillary line, and align the movable arm along the humerus toward the lateral epicondyle. Use the angle in degrees of shoulder flexion from the horizontal as the score.
6. If using the tape measure, use the distance from the lateral epicondyle or the cubital fossa to the surface of the table as the score.

Analysis and Interpretation of Data

With the tape measure, a positive measurement would indicate ability of the shoulder to move past the horizontal position (shoulder hyperflexion), while a negative measurement would indicate that the shoulder is not strictly level with the table and a relative lack of flexibility.

Statistics

Manske and colleagues (2006) found, in comparing normal individuals to collegiate-level pitchers, that the throwers had significantly shorter latissimus dorsi on the dominant side. Intrarater reliability for the upper extremity flexibility assessment was found to be high at 0.83 to 0.814 (intraclass correlation coefficient [ICC] 3,2).

Note

It is important to take care when performing this measurement in a client who indicates a history of glenohumeral dislocation or subluxation.

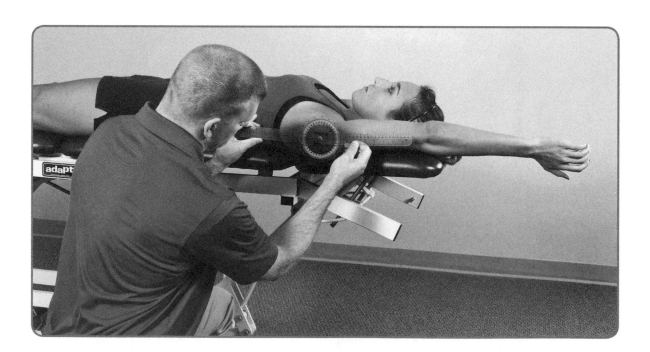

PECTORALIS MAJOR ASSESSMENT

▶ **Purpose:** To determine the extensibility, length, and excursion of the pectoralis major. For the origin, insertion, action, and innervation of these muscles, see table 5.4 on pages 78-79.

▶ **Equipment:** Stable testing surface, standard goniometer or standard tape measure.

Procedure (Reese & Bandy 2002)

1. Position the client so that he or she is supine with hands clasped behind the head.
2. Ensure that the cervical spine is not excessively flexed or extended.
3. Ensure that the client keeps the cervical spine in neutral and hands clasped behind the head.
4. Instruct the client to relax the shoulders, allowing the elbows to move toward the table.
5. Ensure that the lumbar spine remains flat on the table.

6. Use a tape measure to determine the distance in centimeters between the olecranon process of the humerus and the table surface.

Analysis and Interpretation of Data

There are no known normative data for this flexibility measure.

Statistics

Intrarater reliability for upper extremity flexibility assessment was found to be high at 0.83 to 0.814 (ICC 3,2) (Manske et al. 2006).

Note

Be careful when performing this measurement in a client who indicates a history of glenohumeral dislocation or subluxation.

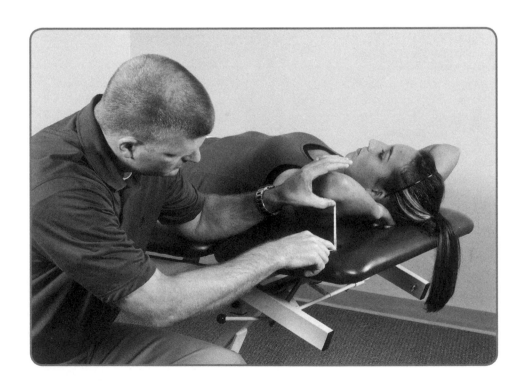

UPPER (CLAVICULAR)—PECTORALIS MAJOR ASSESSMENT

▶ **Purpose:** To determine the extensibility, length, and excursion of the upper (clavicular) pectoralis major. For the origin, insertion, action, and innervation of these muscles, see table 5.4 on pages 78-79.

▶ **Equipment:** Stable testing surface, standard goniometer or standard tape measure.

Procedure (Reese & Bandy 2002)

1. Position the client so that he or she is supine with shoulder laterally rotated and abducted to 90°, elbows fully extended, forearm supinated, and lumbar spine flat against the table surface.
2. Ensure that the client maintains the shoulder in lateral rotation and 90° of abduction and full elbow extension and forearm supination.
3. Ask the client to relax all shoulder muscles, allowing the shoulder to move into maximal horizontal abduction.
4. Ensure that the client keeps the lumbar spine flat against the support surface and that trunk rotation does not occur.

5. Using the tape measure, determine the distance in centimeters between the lateral epicondyle of the humerus and the table.

Analysis and Interpretation of Data

There are no known normative data for this flexibility measure.

Statistics

Intrarater reliability for upper extremity flexibility assessment was found to be high at 0.83 to 0.814 (ICC 3,2) (Manske et al. 2006).

Note

Take care when performing this measurement in a client who indicates a history of glenohumeral dislocation or subluxation.

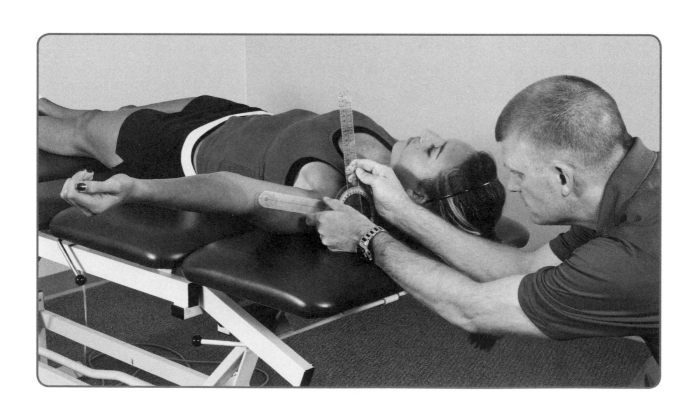

LOWER (STERNAL)—PECTORALIS MAJOR ASSESSMENT

▶ **Purpose:** To determine the extensibility, length, and excursion of the lower (sternal) pectoralis major. For the origin, insertion, action, and innervation of these muscles, see table 5.4 on pages 78-79.

▶ **Equipment:** Stable testing surface, standard goniometer or standard tape measure.

Procedure (Reese & Bandy 2002)

1. Position the client so that he or she is supine with the shoulder laterally rotated and abducted to 135°, elbows fully extended, forearm supinated, and lumbar spine flat against the testing surface.

2. Ensure that the client maintains the shoulder in lateral rotation and 135° of abduction and full elbow extension and forearm supination.

3. Ask the client to relax all shoulder muscles, allowing the shoulder to move into maximal horizontal abduction.

4. Ensure that the client keeps the lumbar spine flat against support surface and does not rotate the trunk.

5. If using the goniometer, measure the angle in degrees of shoulder flexion from the horizontal; place the axis at the shoulder glenohumeral joint, place the stable arm horizontally along the midaxillary line, and align the movable arm along the humerus toward the lateral epicondyle.

6. If using the tape measure, determine the distance in centimeters between the lateral epicondyle of the humerus and the table.

Analysis and Interpretation of Data

There are no known normative data for this flexibility measure.

Statistics

Intrarater reliability for upper extremity flexibility assessment was found to be high at 0.83 to 0.814 (ICC 3,2) (Manske et al. 2006).

Note

Take care when performing this measurement in a client who indicates a history of dislocation or subluxation.

PECTORALIS MINOR ASSESSMENT

▶ **Purpose:** To determine the extensibility, length, and excursion of the pectoralis minor muscle. For the origin, insertion, action, and innervation of these muscles, see table 5.4 on pages 78-79.

▶ **Equipment:** Stable testing surface, plastic ruler or flexible measuring tape.

Procedure (Reese & Bandy 2002)

1. Position the client so that he or she is supine with arms at side, shoulders laterally rotated, forearms supinated, and lumbar spine flat against the support surface.

2. Ensure that the client maintains the correct position.

3. Have the client relax the shoulder muscles, allowing the posterior border of the acromion process to move toward the support surface.

4. Use a tape or plastic ruler to measure the distance in centimeters between the posterior border of the acromion process and the supporting surface.

Analysis and Interpretation of Data

Manske and colleagues (2006) have found, in comparisons of normal subjects to collegiate pitchers, that the nondominant pectoralis minor is statistically shorter than the dominant pectoralis minor in both groups.

Statistics

■ Borstad (2006) reported that this form of measurement for pectoralis minor length demonstrates a low correlation with actual pectoralis muscle length.

■ Manske and colleagues (2006) found intrarater reliability for upper extremity flexibility assessment to be high at 0.83 to 0.814 (ICC 3,2).

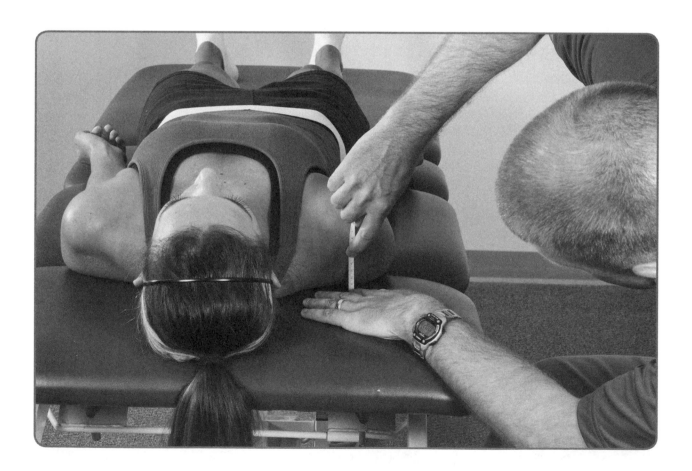

PECTORALIS MINOR ASSESSMENT (BORSTAD METHOD)

▶ **Purpose:** To determine the extensibility, length, and excursion of the pectoralis minor. For the origin, insertion, action, and innervation of these muscles, see table 5.4 on pages 78-79.

▶ **Equipment:** Stable testing surface, flexible tape.

Procedure

1. Position the client so that he or she is supine with arms at side, shoulders laterally rotated, forearms supinated, and lumbar spine flat against the support surface.

2. Ensure that the client maintains the correct position.

3. Have the client relax the shoulder muscles, allowing the posterior border of the acromion process to move toward the support surface.

4. Using a flexible tape, measure the distance in centimeters between the sternal notch and the coracoid process.

Analysis and Interpretation of Data

There are no known normative data for this flexibility measure.

Statistics

Borstad (2006) reports that this form of measurement for pectoralis minor length demonstrates a high correlation with actual pectoralis muscle length.

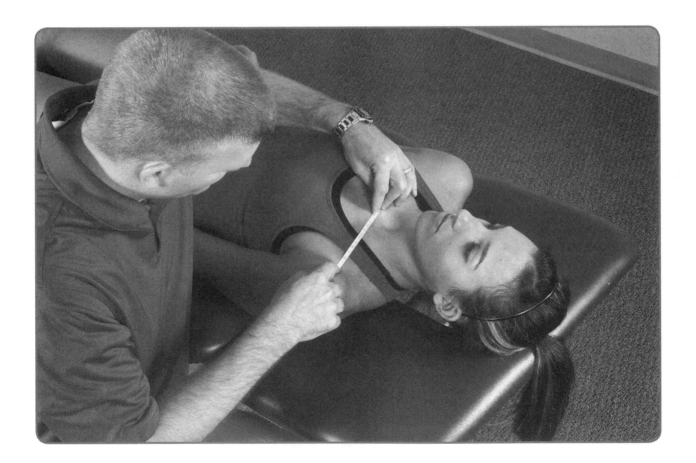

TRICEPS ASSESSMENT

▶ **Purpose:** To determine the extensibility, length, and excursion of the triceps. For the origin, insertion, action, and innervation of this muscle, see table 5.3 on page 78.

▶ **Equipment:** Standard goniometer.

Procedure (Greenman 1996; Reese & Bandy 2002)

1. Position the client so that he or she is seated with shoulder in full flexion, elbow extended, and forearm supinated.
2. Flex the client's elbow through available ROM while maintaining the humerus in full shoulder flexion.

3. To measure the amount of elbow flexion achieved via goniometer, have the axis at the lateral epicondyle, the stationary arm at the midline of the humerus, and the movable arm along the lateral midline of the radius toward the radial styloid process.

Analysis and Interpretation of Data

There are no known normative data for this flexibility measure.

Statistics

There are no known reliability data for this flexibility measure.

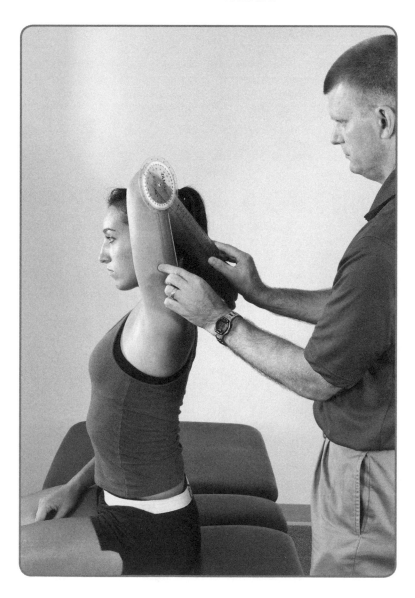

BICEPS ASSESSMENT

▶ **Purpose:** To determine the extensibility, length, and excursion of the biceps. For the origin, insertion, action, and innervation of this muscle, see table 5.3 on page 78.

▶ **Equipment:** Standard goniometer.

Procedure (Greenman 1996; Reese & Bandy 2002)

1. Position the client so that he or she is supine with the shoulder at the edge of the plinth, elbow extended, and forearm pronated.

2. Extend the client's shoulder through available ROM while maintaining elbow in full extension.

3. Measure the amount of shoulder extension via goniometer by placing the lateral epicondyle at the axis of motion; the midline of the thorax or vertical is the landmark for the stationary arm, and the lateral midline of the humerus is the landmark for the movable arm.

Analysis and Interpretation of Data

There are no known normative data for this flexibility measure.

Statistics

There are no known reliability data for this flexibility measure.

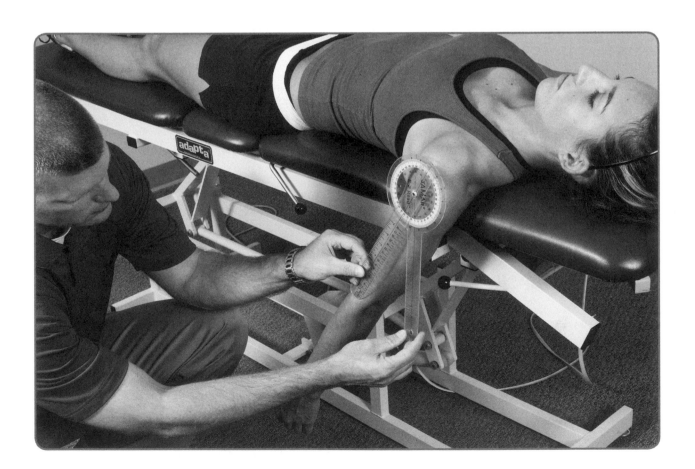

Trunk Assessments

LUMBAR ERECTOR SPINAE ASSESSMENT

▶ **Purpose:** To determine the range of motion of the lumbar erector spinae. For the origin, insertion, action, and innervation of these muscles, see table 5.5 on page 79.

▶ **Equipment:** Stable testing surface, measuring tape.

Procedure (Janda 1983)

1. Position the client so that he or she is long sitting on a testing surface, keeping the pelvis as vertical as possible.
2. Stand to the side of the client and monitor the movement.
3. Ask the client to forward bend at the lumbopelvic region and move the forehead toward the knees (first position).
4. For the second part of the assessment, have the client sit at the end of the table with bilateral knees flexed.

5. Ask the client to forward bend as far as possible, moving the forehead toward the knees without moving the pelvis (second position).

Analysis and Interpretation of Data

■ In the first position, posterior tilting of the pelvis is a sign of adaptive shortening of the hamstrings.

■ In the first position, an adult should achieve a distance of 10 cm or less between the forehead and knees, as well as demonstrating a smooth, even curve of the spine.

■ If the forward bending of the trunk is greater in the second position than in the first position, this is usually the result of an increased tilt of the pelvis and adaptive shortening of the hamstrings versus adaptive shortening of the erector spinae.

Statistics

There are no known reliability data for this flexibility measure.

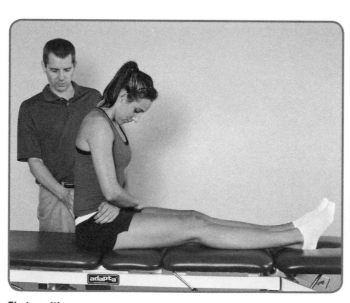

First position.

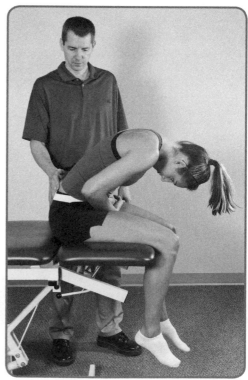

Second position.

QUADRATUS LUMBORUM ASSESSMENT I

▶ **Purpose:** To determine the extensibility, length, and excursion of the quadratus lumborum. For the origin, insertion, action, and innervation of this muscle, see table 5.6 on page 79.

▶ **Equipment:** Stable testing surface, tape measure.

Procedure (Greenman 1996)

1. Position the client so that he or she is lying on the nontest side with the bottom leg bent slightly at the hip and knee for stability.

2. Stand in front of the client and monitor the quadratus lumborum muscle and iliac crest.

3. Have the client extend the top leg beyond the edge of the table and allow the top (test) leg to drop toward the floor.

Analysis and Interpretation of Data

■ Measure the distance from the epicondyle of the medial femoral condyle of the knee to the floor and compare to the other side. Shortness and tightness are present if the ROM is asymmetric.

■ Also assess for the development of tension in quadratus lumborum, as well as for reduced caudad motion of the iliac crest being assessed.

■ There are no known normative data for this flexibility measure.

Statistics

There are no known reliability data for this flexibility measure.

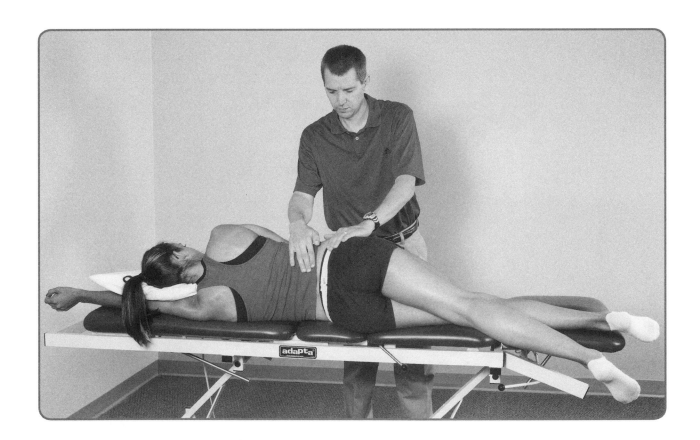

QUADRATUS LUMBORUM ASSESSMENT II

▶ **Purpose:** To determine the extensibility, length, and excursion of the quadratus lumborum. For the origin, insertion, action, and innervation of this muscle, see table 5.6 on page 79.

▶ **Equipment:** Stable testing surface.

Procedure (Greenman 1996; Janda 1983)

1. Position the client so that he or she is side-lying with the hips and knees flexed at about 45°.

2. Monitor from behind the client. (Photo shows clinician in front of client so that view of client position is unobstructed.)

3. Ask the client to push up sideways from the table to a point where the pelvis begins to move.

4. Monitor to avoid trunk flexion or rotation during the assessment.

Analysis and Interpretation of Data

◼ Tightness can be less accurately assessed during lumbar side-bending to the contralateral side in standing. Normal movement would be smooth, symmetrical curving of the spine in both directions.

◼ Shortness and tightness are present if the ROM is asymmetric.

Statistics

There are no known reliability data for this flexibility measure.

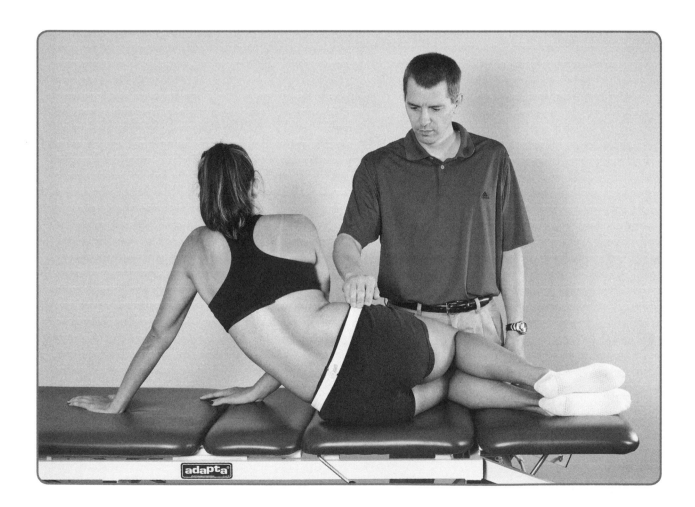

LATISSIMUS DORSI ASSESSMENT

▶ **Purpose:** To determine the extensibility, length, and excursion of the latissimus dorsi. For the origin, insertion, action, and innervation of these muscles, see table 5.4 on pages 78-79.

▶ **Equipment:** Stable testing surface, standard goniometer.

Procedure (Lee 2004)

1. Position the client so that he or she is initially sitting in a neutral spine position, arms resting at the side.

2. Monitor from behind the client during the entire assessment.

3. Instruct the client to rotate the trunk to the left (first position) and then to the right.

4. Note the quantity and quality of the motion through the thoracic and lumbar spine.

5. Instruct the client to flex the upper extremities to 90°, fully externally rotate and adduct the shoulders, and approximate the hypothenar eminences of bilateral upper extremities.

6. Instruct the client to rotate the trunk to the left (second position) and to the right.

7. Again observe the quantity and quality of the motion through the thoracic and lumbar spine.

Analysis and Interpretation of Data

■ Comparison is made of quantity and quality of motion with the arms at the side in bilateral directions, as well as with the arms elevated, externally rotated, and adducted.

■ Comparisons are of side-to-side differences, as well as between arms at the side and arms when elevated.

■ The motion is markedly reduced in the second position (figure 5.17b) when the latissimus dorsi muscle is tight because this position increases the tension through this muscle.

Statistics

There are no known reliability data for this flexibility measure.

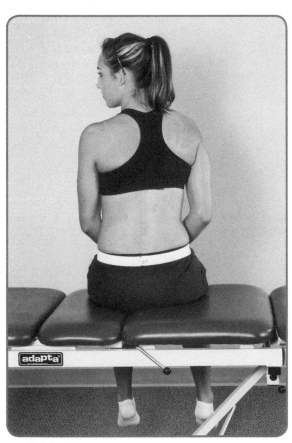

First position.

Second position.

Lower Extremity Assessments

HIP EXTENSORS ASSESSMENT

▶ **Purpose:** To determine extensibility, length, and excursion of the hip extensors. For the origin, insertion, action, and innervation of these muscles, see table 5.7 on page 79.

▶ **Equipment:** Stable testing surface, measurement device (goniometer, bubble goniometer, etc.).

Procedure (Bullock-Saxton & Bullock 1994)

1. Position the client so that he or she is supine on a stable surface.

2. Passively bring the hip into flexion while palpating the posterior superior iliac spine (PSIS) on the same side.

3. Take the measurement according to standard procedure for hip flexion-extension with the stationary arm of the goniometer parallel to the trunk and the movable arm parallel to the thigh.

4. The axis of motion is the greater trochanter.

5. As soon as the PSIS moves posteriorly, stop the movement and record the measurement.

Analysis and Interpretation of Data

Tightness is present if the ROM is asymmetric.

Statistics

Interexaminer reliability (ICC) = .87.

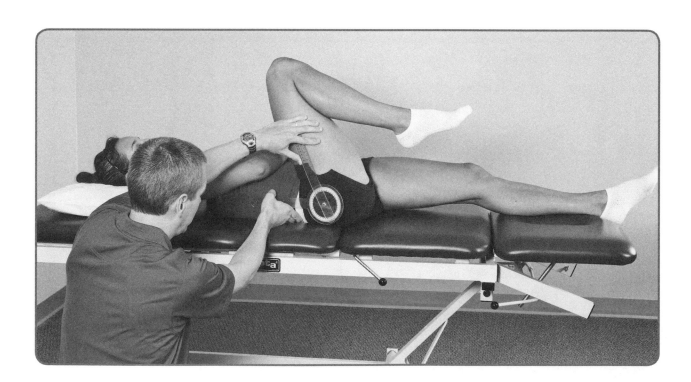

HIP FLEXORS ASSESSMENT

▶ **Purpose:** To determine the extensibility, length, and excursion of the hip flexors. For the origin, insertion, action, and innervation of these muscles, see table 5.8 on page 80.

▶ **Equipment:** Stable testing surface, measurement device (goniometer, bubble goniometer, etc.).

Procedure (Bullock-Saxton & Bullock 1994)

1. Position the client so that he or she is supine on a stable surface with both lower extremities over the edge.
2. Passively flex both hips; then slowly lower the side being tested.

3. Take the measurement according to standard procedure for hip flexion-extension with the stationary arm of the goniometer parallel to the trunk and the movable arm parallel to the thigh.
4. The axis of motion is the greater trochanter.
5. Record the measurement when the limb ceases to move.

Analysis and Interpretation of Data

Tightness is present if the ROM is asymmetric.

Statistics

Interexaminer reliability (ICC) = .98.

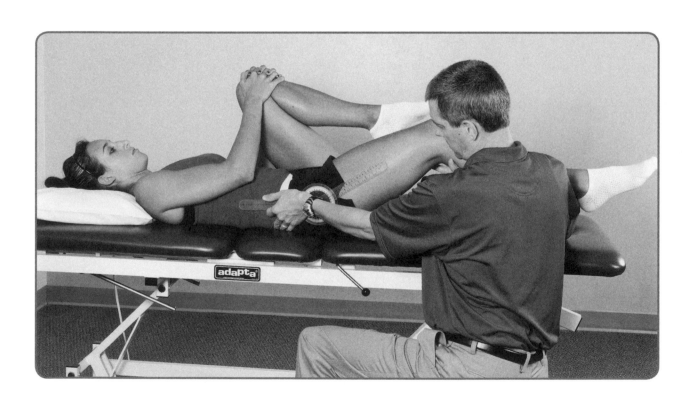

EXTERNAL ROTATORS OF THE HIP ASSESSMENT

▶ **Purpose:** To determine the extensibility, length, and excursion of the external rotators of the hip. For the origin, insertion, action, and innervation of these muscles, see table 5.9 on page 81.

▶ **Equipment:** Stable testing surface, measurement device (goniometer, bubble goniometer, etc.).

Procedure (Bullock-Saxton & Bullock 1994)

1. Position the client so that he or she is prone on the testing surface.
2. Passively flex the knee to 90°.
3. Have the stationary arm of the goniometer running parallel to the testing surface and the movable arm running parallel to the tibia of the testing limb.
4. Palpate the opposite PSIS and passively internally rotate the limb being tested.
5. Take the measurement when you note rotation of the pelvis.

Analysis and Interpretation of Data

Tightness is present if the ROM is asymmetric.

Statistics:

Interexaminer reliability (ICC) = .99.

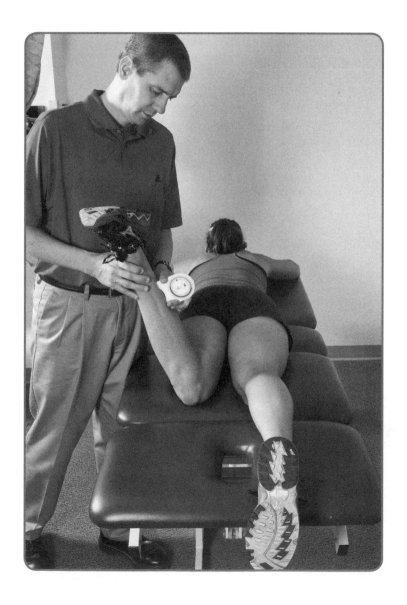

INTERNAL ROTATORS OF THE HIP ASSESSMENT

▶ **Purpose:** To determine the extensibility, length, and excursion of the internal rotators of the hip. For the origin, insertion, action, and innervation of these muscles, see table 5.10 on page 81.

▶ **Equipment:** Stable testing surface, measurement device (goniometer, bubble goniometer, etc.).

Procedure (Bullock-Saxton & Bullock 1994)

1. Position the client so that he or she is prone on testing surface.

2. Passively flex the knee to 90°.

3. Have the stationary arm of the goniometer running parallel to the testing surface and the movable arm running parallel to the tibia of the testing limb.

4. Palpate the opposite PSIS and passively externally rotate the limb being tested.

5. Take the measurement when you note rotation of the pelvis.

Analysis and Interpretation of Data

Tightness is present if the ROM is asymmetric.

Statistics

Interexaminer reliability (ICC) = .98.

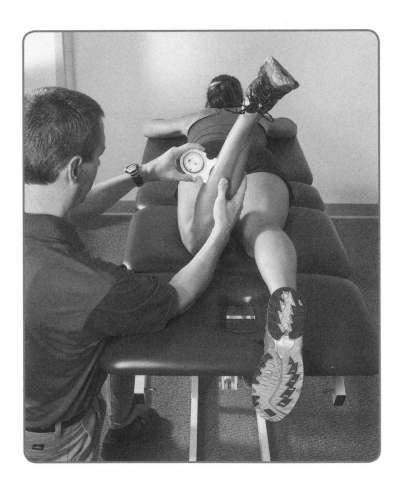

PSOAS ASSESSMENT I

▶ **Purpose:** To determine the extensibility, length, and excursion of the psoas. For the origin, insertion, action, and innervation of these muscles, see table 5.8 on page 80.

▶ **Equipment:** Ruler or measuring stick.

Procedure (Greenman 1996)

1. Position the client so that he or she is prone with upper trunk relaxed.
2. Lift the client's leg with the knee flexed while blocking pelvic motion at the ischial tuberosity until engaging the barrier.

Analysis and Interpretation of Data

■ Normally the knee can be lifted 6 in. (15 cm) off the table. If this distance is less than 6 in., tightness and shortness of the psoas are present.

■ Shortness and tightness are present if the ROM is asymmetric.

Statistics

There are no known reliability data for this flexibility measure.

Notes

■ Pay careful attention to stabilization of the pelvis and avoiding anterior pelvic tilt.

■ This position can also stretch the anterior capsule of the hip.

■ Flexing the knee from this position can enable you to check for rectus femoris muscle length, but pay careful attention to femoral nerve stretch and similar symptoms.

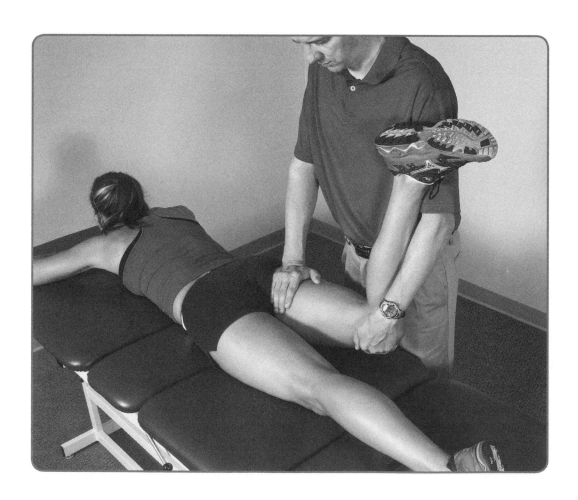

PSOAS ASSESSMENT II

▶ **Purpose:** To determine the extensibility, length, and excursion of the psoas. For the origin, insertion, action, and innervation of these muscles, see table 5.8 on page 80.

▶ **Equipment:** Stable testing surface, standard goniometer.

Procedure (Greenman 1996)

1. Position the client so that he or she is supine at the end of the table.
2. Have the client hold the nontest leg in a flexed position.
3. Help the client to extend the leg that is being tested.
4. Have the stationary arm of the goniometer lined up parallel with the trunk; the axis is at the midhip joint, and the moving arm runs parallel to the thigh using the lateral epicondyle of the knee as a reference point.

Analysis and Interpretation of Data

■ If the back of the thigh does not contact the table, psoas shortness is present (Greenman 1996).

■ If a hip flexion angle of at least 7° below horizontal is not achieved, the iliopsoas is considered tight (Harvey & Mansfield 2000).

Statistics

There are no known reliability data for this flexibility measure.

Notes

■ You can use your proximal hand to monitor lumbar spine position.

■ The lumbar spine should maintain contact with table.

■ The client can compensate with increased lumbar lordosis or anterior pelvic tilt.

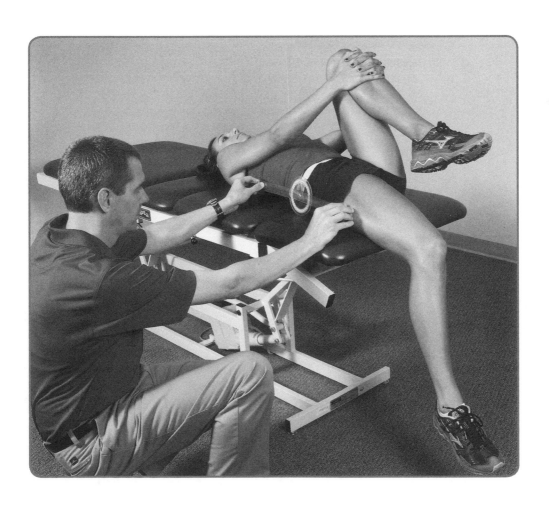

RECTUS FEMORIS ASSESSMENT

▶ **Purpose:** To determine the extensibility, length, and excursion of the rectus femoris. For the origin, insertion, action, and innervation of this muscle, see table 5.8 on page 80.

▶ **Equipment:** Stable testing surface, standard goniometer.

Procedure (Greenman 1996; Kendall et al. 2005)

1. Position the client so that he or she is supine with buttocks on the end of the table. Stand at the end of the table.

2. Have the client hold the nontest leg in a flexed position, with the thigh of the test leg on the table.

3. Monitor that the client keeps the nontest knee in the initial flexed position.

4. Assess flexion ROM at the knee.

5. Use the following goniometric reference points: The stationary arm is parallel to the thigh and femur and lined up with the midhip joint; the axis is at midknee; and the moving arm reference point is the lateral malleolus of the ankle as it runs parallel to midtibia.

6. You can use your proximal hand to monitor lumbar spine position.

7. Make sure that the lumbar spine maintains contact with the table to avoid anterior pelvic tilt.

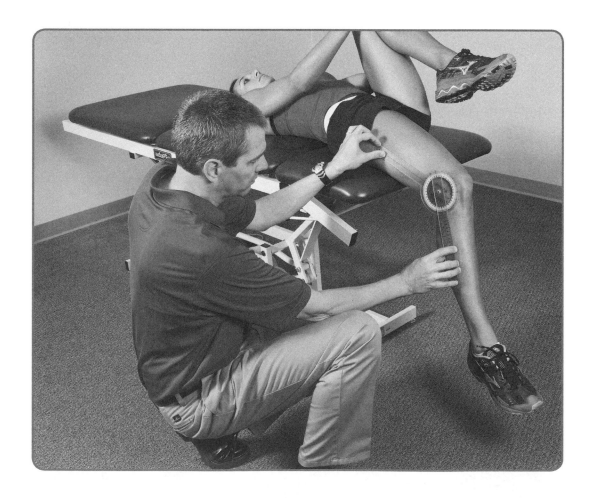

Analysis and Interpretation of Data

The following are guidelines to use in order to determine which specific structure is at fault:

■ Less than 10° to 15° of hip extension indicates a tight iliopsoas. Simultaneous extension of the knee during this maneuver indicates tightness of the rectus femoris (Greenman 1996).

■ Knee flexion: If less than 100° to 105° is available, the rectus femoris is tight (Greenman 1996).

■ With low back and sacrum flat on the table, the posterior thigh touches the table and the knee flexes approximately 80° (Kendall et al. 2005).

■ Shortness in the tensor fasciae latae is indicated by abduction of the thigh as the hip extends, by extension of the knee if the thigh is prevented from abducting or is passively adducted as the hip is extended, or by internal rotation of the thigh (Kendall et al. 2005).

Statistics

There are no known reliability data for this flexibility measure.

TENSOR FASCIAE LATAE ASSESSMENT

▶ **Purpose:** To determine the extensibility, length, and excursion of the tensor fasciae latae (TFL). For the origin, insertion, action, and innervation of this muscle, see table 5.10 on page 81.

▶ **Equipment:** Stable testing surface, standard goniometer.

Procedure (Greenman 1996)

1. Position the client so that he or she is supine with buttocks on end of the table. Stand at the end of the table.

2. Have the client hold the nontest leg in the flexed position with the thigh of the test leg on the table.

3. Introduce internal rotation and slight adduction of the thigh and external rotation of the tibia, sensing for resistance, and monitor the contour of the lateral thigh.

4. Goniometric measurement is with standard hip adduction measurement: axis over ipsilateral ASIS, stationary arm lined with contralateral ASIS and movable arm lined with mid patella.

Analysis and Interpretation of Data

▩ If resistance is encountered and the lateral thigh shows a groove, tightness of the TFL is present.

▩ Hip adduction: If less than 15° to 20° is available, the tensor fasciae latae and iliotibial (IT) band are tight (Greenman 1996).

Statistics

There are no known reliability data for this flexibility measure.

Notes

▩ You can use your proximal hand to monitor lumbar spine position.

▩ The lumbar spine should maintain contact with the table to avoid anterior pelvic tilt.

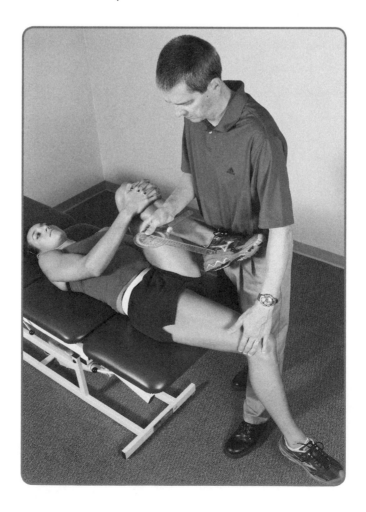

HIP ADDUCTOR ASSESSMENT (LONG VERSUS SHORT)

▶ **Purpose:** To determine the extensibility, length, and excursion of the hip adductor. For the origin, insertion, action, and innervation of these muscles, see table 5.11 on page 82.

▶ **Equipment:** Stable testing surface, standard goniometer.

Procedure (Greenman 1996)

1. Position the client so that he or she is supine with the leg to be tested close to the edge of the table.
2. Instruct the client to abduct the nontest leg 15° to 20° at the hip, with the heel over the edge of the table.
3. Stand at the client's side, next to the test leg.
4. Maintaining the leg to be assessed in full extension, passively abduct it (normal range is 40°).
5. When the full ROM is reached, passively flex the knee of the test leg and attempt to abduct the leg further.

6. Goniometric measurement is as with standard hip abduction technique.

Analysis and Interpretation of Data

■ Hip abduction: If less than 15° to 20° is available, the short hip adductors are tight (Greenman 1996).

■ If the maximum hip abduction range does not increase when the knee is flexed, the single joint adductors (pectineus, adductor magnus, adductor longus, and adductor brevis) are shortened.

■ If the hip abduction range does increase with the knee passively flexed, the double joint adductors (gracilis, semimembranosus, and semitendinosus) are shortened.

Statistics

There are no known reliability data for this flexibility measure.

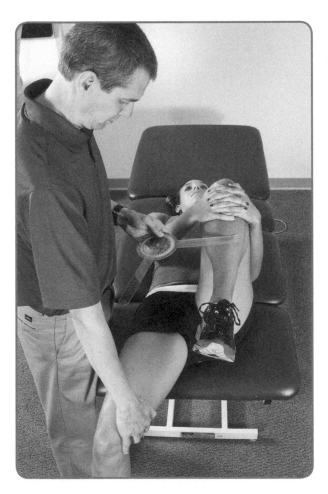

OBER'S TEST—ILIOTIBIAL BAND ASSESSMENT

▶ **Purpose:** To determine the extensibility, length, and excursion of the iliotibial band.

▶ **Equipment:** Stable testing surface, standard goniometer or cloth tape measure.

Procedure

1. Position the client so that he or she is side-lying with the test side up.

2. Have the client bend the downside leg to allow for a more stable base to prevent trunk and total-body rotation.

3. Ensure that alignment is in the frontal plane throughout the body, not allowing for rotation or hip flexion compensation.

4. Maintain the client in frontal plane alignment, not allowing rotation at the trunk.

5. Stabilize the ipsilateral hip and hold the client's leg with the caudal hand on the medial surface of the knee.

6. Flex the client's knee to 20° and slightly externally rotate the hip to bring the IT band over the greater trochanter.

7. Also abduct the hip slightly prior to the initiation of the testing movement.

8. While maintaining the cranial hand on the hip for stabilization, gradually lower (or adduct) the client's thigh to end range (typically indicated by iliac crest movement).

Analysis and Interpretation of Data

■ You can measure distance from the medial epicondyle of the knee to the table or measure the angle from horizontal for objective measurement.

■ According to Kendall and colleagues (2005), the measurement is normal when the thigh drops to 10° below the horizontal.

■ Anything above 10° is indicative of IT band tightness.

Statistics

Reliability: ICC (goniometer) from 0.82 to 0.92 (Gajdosik et al. 2003); ICC (inclinometer) = 0.94 (Melchione & Sullivan 1993); ICC = 0.91 (Reese & Bandy 2003).

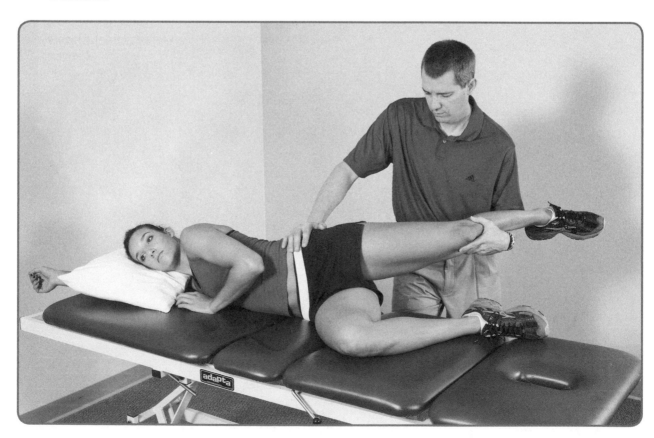

PIRIFORMIS ASSESSMENT

▶ **Purpose:** To determine the extensibility, length, and excursion of the piriformis. For the origin, insertion, action, and innervation of this muscle, see table 5.8 on page 80.

▶ **Equipment:** Stable testing surface, standard goniometer.

Procedure (Kendall et al. 2005; Kroon & Kruchowsky 2006)

1. Position the client so that he or she is supine on a stable testing surface with bilateral lower extremities extended.
2. Passively flex the client's hip to 90°, then adduct and internally rotate to the end range of the movement.

Analysis and Interpretation of Data

▪ For normal length, the hip should be able to flex 90°, adduct approximately 20° (knee should reach midsagittal plane), and internally rotate 20°.

▪ A tight or short piriformis is indicated if the client is unable to achieve normal range of movement.

Statistics

There are no known reliability data for this flexibility measure.

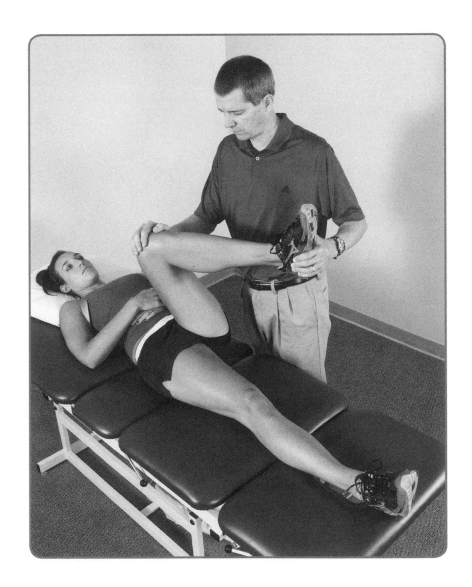

PIRIFORMIS (FAIR TEST)

▶ **Purpose:** To determine the extensibility, length, and excursion of the piriformis. For the origin, insertion, action, and innervation of this muscle, see table 5.8 on page 80.

▶ **Equipment:** Stable testing surface.

Procedure (Fishman et al. 2002)

1. Position the client so that he or she is side-lying on the nontest side with the hip and knee slightly bent for stability.

2. Have the client's trunk in normal postural alignment.

3. Stand directly behind the client at the level of the hips.

4. Bring the test leg into a position of flexion, adduction, and internal rotation (FAIR).

Analysis and Interpretation of Data

■ This test can be used to assess for flexibility of piriformis in the absence of symptoms (compare to other side).

■ It can also be used to assess for piriformis muscle pain involvement; the test is considered positive if pain is elicited at the intersection of the sciatic nerve and the piriformis.

Statistics

For piriformis dysfunction: sensitivity .88; specificity .83; +LR 5.2, −LR .14.

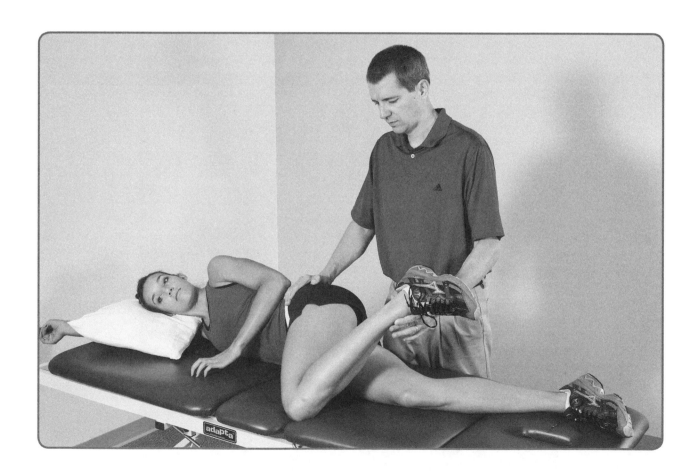

HIP FLEXIBILITY: FIGURE FOUR TEST

▶ **Purpose:** To determine the extensibility, length, and excursion of the hip.

▶ **Equipment:** Stable testing surface, cloth tape measure.

Procedure (Harvey & Mansfield 2000)

1. Position the client so that he or she is prone on the testing surface and places the sole of the foot on the inside of the opposite knee.

2. Instruct the client to attempt to push the hip to the floor using the gluteals.

3. Instruct the client to maintain the medial aspect of the knee and ankle in contact with the surface throughout the test.

4. Measure the distance from the anterior superior iliac spine (ASIS) to the testing surface with the tape measure.

5. Note any reports of pain or discomfort from the client.

Analysis and Interpretation of Data

■ A distance of more than 3 cm indicates tightness of the hip joint anterior capsule (hip musculature and contralateral sacroiliac joint) (Harvey & Mansfield 2000).

■ An excessively high ASIS is indicated by decreased hip external rotation, extension, or both.

■ This may be the result of a tight TFL or rectus femoris, iliopsoas, or capsular restriction (Kroon & Kruchowsky 2006).

Statistics

There are no known reliability data for this flexibility measure.

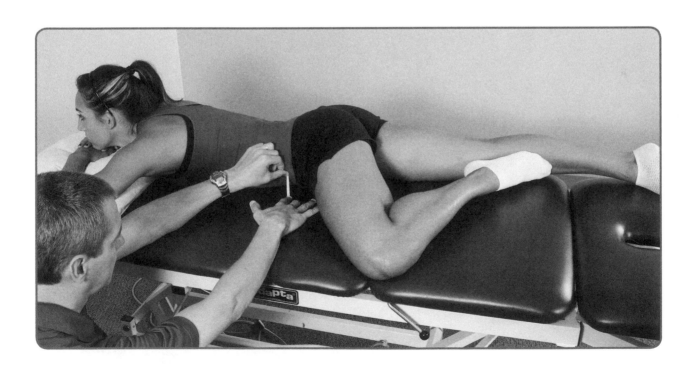

HAMSTRING LENGTH: ACTIVE SUPINE 90/90 POSITION ASSESSMENT

▶ **Purpose:** To determine the extensibility, length, and excursion of the hamstrings. For the origin, insertion, action, and innervation of this muscle, see table 5.12 on page 82.

▶ **Equipment:** Stable testing surface, standard goniometer.

Procedure

1. Position the client in supine, with hip flexed to 90°.
2. Place the contralateral lower extremity on the support surface with the knee fully extended.
3. It is imperative that the contralateral lower extremity be maintained in this position throughout testing.
4. Instruct the client to actively extend the knee through the full available ROM until firm muscular resistance is felt while the hip is maintained in 90° of flexion.
5. Align the goniometer with the stationary arm along the femur with the reference point of the greater trochanter of the femur; the axis of movement is the lateral epicondyle at the knee, and the moving arm is aligned with the lateral malleolus.
6. Use the measured angle as the score.

Analysis and Interpretation of Data

There are no known normative data for this flexibility measure.

Statistics

Intratester reliability data for this test are given in table 5.13 on page 83.

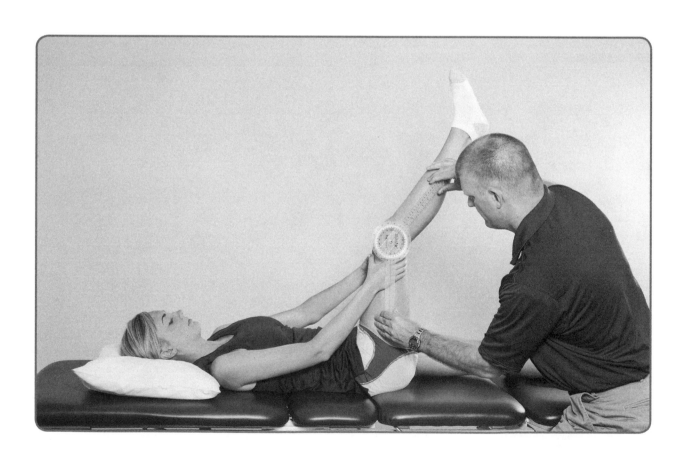

HAMSTRINGS: PASSIVE STRAIGHT LEG RAISE METHOD

▶ **Purpose:** To determine the extensibility, length, and excursion of the hamstrings. For the origin, insertion, action, and innervation of this muscle, see table 5.12 on page 82.

▶ **Equipment:** Stable testing surface, standard goniometer.

Procedure (Greenman 1996)

1. Position the client so that he or she is supine on the table.
2. Stand on the same side of the table as the leg being tested.
3. Monitor the opposite ASIS while lifting the test leg to the barrier of hip flexion.
4. Compare with the opposite side.
5. Align the goniometer such that the stationary arm is aligned parallel to the trunk, the axis is through the midhip-acetabulum, and the moving arm is parallel to the midthigh with the lateral malleolus as reference alignment.
6. You can perform the test with the leg adducted and abducted to test the difference in tightness of the medial and lateral hamstrings.
7. Monitor for signs and symptoms of neural tension of the lower extremity (straight leg raise test).
8. Use the angle measured with the goniometer as the score.

Analysis and Interpretation of Data

If the sides are asymmetric, shortness and tightness are present in the involved hamstring muscles.

Statistics

Intratester reliability data for this test are given in table 5.14 on page 83.

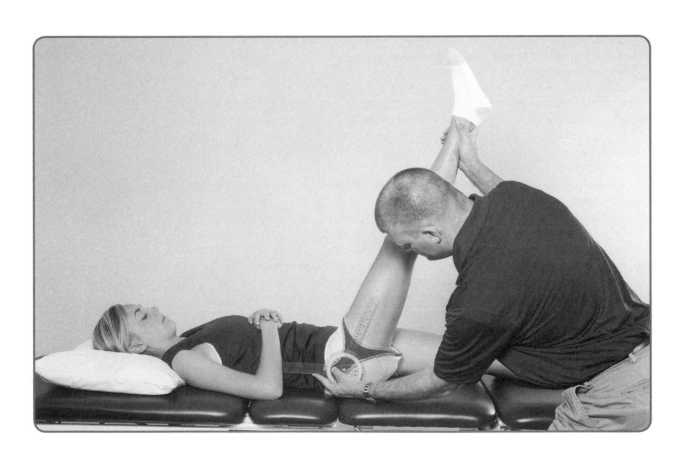

HAMSTRING LENGTH: PASSIVE SUPINE 90/90 POSITION ASSESSMENT

▶ **Purpose:** To determine the extensibility, length, and excursion of the hamstrings. For the origin, insertion, action, and innervation of this muscle, see table 5.12 on page 82.

▶ **Equipment:** Stable testing surface, standard goniometer.

Procedure (Reese & Bandy 2002)

1. Position the client so that he or she is supine, with hip flexed to 90°.
2. Place the contralateral lower extremity on the support surface with knee fully extended.
3. It is imperative that the contralateral lower extremity be maintained in this position throughout testing.
4. After instructing client on the motion desired, extend the client's knee through full available ROM passively until firm muscular resistance is felt while maintaining the hip in 90° of flexion.
5. Align the goniometer with the stationary arm along the femur (or inclinometer along mid shin as shown), with the greater trochanter of the femur as the reference point; the axis of movement is the lateral epicondyle at the knee, and the moving arm is aligned with the lateral malleolus.
6. Use the angle measured as the score.

Analysis and Interpretation of Data

Youdas and colleagues (2005) have assessed hamstring muscle length in both males and females and have determined normative data for those in the age range of 20 to 80 years. Results indicate that males have a popliteal angle ranging from 138.1° to 142.8° with an average of 141.4°. Women of the same age have popliteal angles that range from 148.7° to 154.8° with an average of 152.0°. These data seem to indicate that women have greater hamstring length at all ages than male counterparts.

Statistics

Intratester reliability data for this test are given in table 5.15 on page 83.

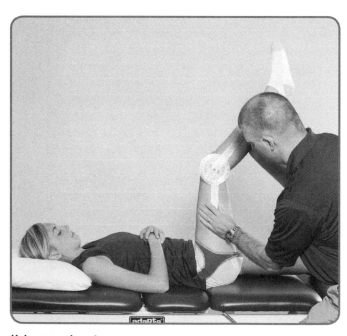

Using a goniometer.

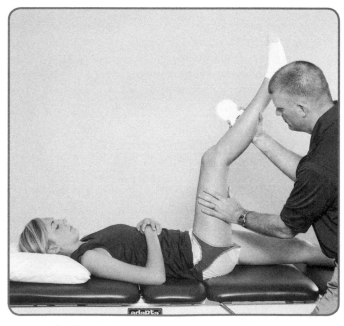

Using an inclinometer.

GASTROCNEMIUS ASSESSMENT

▶ **Purpose:** To determine the extensibility, length, and excursion of the gastrocnemius. For the origin, insertion, action, and innervation of this muscle, see table 5.16 on page 84.

▶ **Equipment:** Stable testing surface, standard goniometer.

Procedure (Reese & Bandy 2002)

1. Position the client so that he or she is supine, with hip and knee extended.
2. Dorsiflex the client's ankle through full available ROM while maintaining the knee in full extension.
3. Maintaining full extension, perform passive dorsiflexion of the ankle.
4. Palpate the following landmarks and align the goniometer accordingly: The stationary arm is at the head of the fibula, the axis is at the lateral malleolus, and the moving arm is parallel to the fifth metatarsal.
5. Ensure that the knee remains in full extension during dorsiflexion.
6. Maintain subtalar neutral; this is thought to allow for pure dorsiflexion by avoiding pronation and supination.
7. Use the goniometric measurement as the score.

Analysis and Interpretation of Data

There are no known normative data for this flexibility measure.

Statistics

Wang and colleagues (1993) examined lower extremity flexibility in long distance runners and found that intratester reliability of gastrocnemius measurements for 10 subjects, in supine, was 0.98 for both the dominant and nondominant lower extremities.

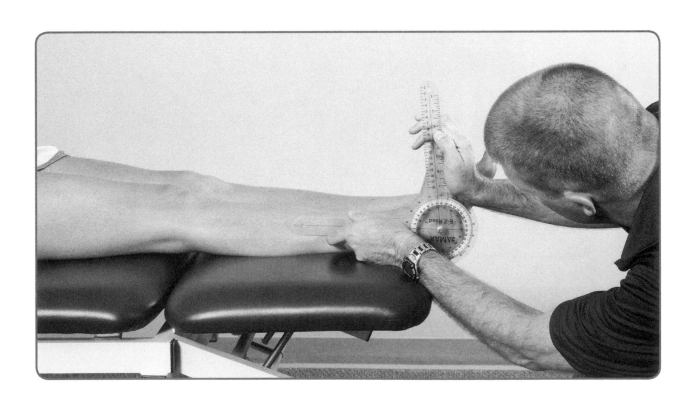

SOLEUS MUSCLE LENGTH TEST: SUPINE

▶ **Purpose:** To determine the extensibility, length, and excursion of the soleus muscle. For the origin, insertion, action, and innervation of this muscle, see table 5.16 on page 84.

▶ **Equipment:** Stable testing surface, standard goniometer.

Procedure (Reese & Bandy 2002)

1. Position the client so that he or she is supine, with hip and knee flexed to about 45°.

2. Place the opposite lower extremity on the support surface with knee fully extended.

3. Dorsiflex the client's ankle through full available ROM while maintaining the hip and knee in 45° of flexion.

4. Maintaining hip and knee in 45° of flexion, perform passive dorsiflexion of the ankle.

5. Palpate the following landmarks and align the goniometer accordingly: The stationary arm is at the head of the fibula, the axis is at the lateral malleolus, and the moving arm is parallel to the fifth metatarsal.

6. Maintain subtalar neutral; this is thought to allow for pure dorsiflexion by avoiding pronation and supination.

7. You can also measure soleus in the prone position with the client's knee bent to 90° of flexion and the same goniometric alignment.

8. Use the goniometric measurement as the score.

Analysis and Interpretation of Data

There are no known normative data for this flexibility measure.

Statistics

Wang and colleagues (1993) examined lower extremity flexibility in long distance runners and found that intratester reliability of soleus measurements in 10 subjects, in prone position, was 0.93 for the dominant limb and 0.94 for the nondominant limb.

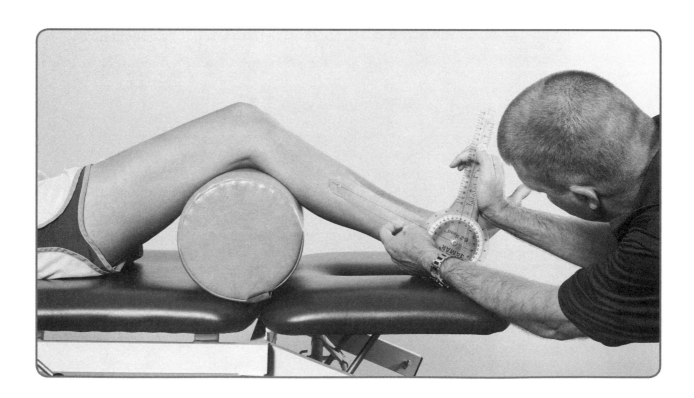

TABLE 5.1

Neck Muscles

Muscle	Origin	Insertion	Action	Innervation
Levator scapulae	Transverse processes of C1-C4	Superomedial border of scapula	Elevation of scapula, scapular adduction and downward rotation	Dorsal scapular nerve C3-C5
Trapezius	Superior nuchal line, occipital protuberance, nuchal ligament	Lateral clavicle, acromion, spine of scapula	Elevation of scapula	Spinal root of accessory nerve C3-C4
Sternocleidomastoid	Mastoid process and lateral superior nuchal line	Sternal head: anterior manubrium Clavicular head: superior medial clavicle	Neck flexion, ipsilateral side-bending, and contralateral rotation	Spinal root of accessory nerve C2-C3
Anterior scalene	Anterior tubercles of transverse processes of C4-C6	1st rib (scalene tubercle)	Elevation of 1st rib, ipsilateral side-bending, and contralateral rotation	C4-C6
Middle scalene	Transverse processes of C2-C7	1st rib	Elevation of 1st rib, ipsilateral side-bending, and contralateral rotation	Ventral rami of cervical spinal nerves C3-C8
Posterior scalene	Transverse processes of C5-C6	2nd rib	Elevation of 2nd rib, ipsilateral side-bending, and contralateral rotation	Ventral rami of cervical spinal nerves C3, C4 C6-C8

TABLE 5.2

Suboccipital Muscles

Muscle	Origin	Insertion	Action	Innervation
Obliquus capitis inferior	C2 spinous process	C1 transverse process	Rotation of atlas (turning head to same side)	Dorsal ramus of C1
Obliquus capitis superior	C1 transverse process	Occipital bone	Extension and lateral flexion of head	Dorsal ramus of C1
Rectus capitis posterior major	C2 spinous process	Occipital bone	Extension of head, rotation of head to same side	Dorsal ramus of C1
Rectus capitis posterior minor	Posterior tubercle of C2	Occipital bone	Extension of head	Dorsal ramus of C1

TABLE 5.3

Upper Arms Muscles

Muscle	Origin	Insertion	Action	Innervation
Biceps brachii	Short head—tip of coracoid process of scapula; long head—supraglenoid tubercle of scapula	Radial tuberosity and bicipital aponeurosis into fascia of forearm	Flexion and supination of forearm; flexion of the arm	Musculocutaneous nerve
Coraco-brachialis	Coracoid process of scapula	Anteromedial surface of midshaft of humerus	Flexion and adduction of arm	Musculocutaneous nerve
Brachilais	Lower half of anterior surface of humerus; intermuscular septa	Ulnar tuberosity	Flexion and adduction of arm	Musculocutaneous nerve (lateral side may receive fibers from radial nerve)
Triceps brachii	Long head—infraglenoid tubercle of scapula; lateral head—posterior surface of humerus above and lateral to groove of radial nerve and lateral intermuscular septum; medial head—posterior surface of humerus below and medial to groove of radial nerve and both intermuscular septa	Proximal end of olecranon of ulna	Extension of forearm; extension of arm (long head)	Radial nerve
Anconeus	Lateral epicondyle of humerus	Lateral side of olecranon of ulna	Extension of forearm	Radial nerve

TABLE 5.4

Upper Extremity Muscles

Muscle	Origin	Insertion	Action	Innervation
Latissimus dorsi	Spinous processes of lower 6 thoracic and all lumbar and sacral vertebrae; posterior part of iliac crest	Medial lip (crest of lesser tubercle) and floor of intertubercular groove of humerus)	Extension, adduction, and medial rotation of arm	Thoracodorsal nerve
Rhomboid minor	Lower part of ligamentum nuchae; spinous processes of 7th cervical and 1st thoracic vertebrae	Medial border of scapula at base of spine	Elevation and retraction of scapula; downward rotation of glenoid cavity	Dorsal scapular nerve
Rhomboid major	Spinous processes of 2nd to 5th thoracic vertebrae	Medial border of scapula below rhomboid minor	Elevation and retraction of scapula; downward rotation of glenoid cavity	Dorsal scapular nerve
Serratus anterior	Ribs 1-8 on anterolateral thoracic wall	Medial border of scapula; heaviest insertion to inferior angle	Protraction of scapula; upward rotation of glenoid cavity; holds medial border against thoracic wall	Long thoracic nerve
Deltoid	Lateral 3rd of clavicle; acromion; spine of scapula	Deltoid tuberosity on shaft of humerus	Abduction (middle); flexion and medial rotation (anterior); extension and lateral rotation (posterior)	Axillary nerve
Supra-spinous	Supraspinous fossa of scapula	Greater tubercle	Abduction of the arm	Suprascapular nerve

Muscle	Origin	Insertion	Action	Innervation
Infra-spinatus	Infraspinous fossa of scapula	Greater tubercle of humerus below supraspinatus	Lateral rotation of arm	Suprascapular nerve
Teres minor	Upper 2/3 of lateral border of scapula	Greater tubercle of humerus below infraspinatus	Lateral rotation of arm	Axillary nerve
Teres major	Inferior angle of scapula	Medial tip of intertubercular groove of humerus	Adduction, medial rotation, and extension of arm	Lower subscapular nerve

TABLE 5.5

Lumbar Erector Spinae Muscles

Muscle	Origin	Insertion	Action	Innervation
Erector spinae Iliocostalis lumborum	Iliac crest, sacrum	Lower borders of lowest seven ribs	Bilateral action: extension of trunk	Dorsal rami of spinal nerves in corresponding area
Longissimus thoracis	Intermediate part of extensor aponeurosis	Lower 10 ribs and adjacent vertebral transverse processes	Unilateral action: lateral trunk flexion ipsilaterally	

TABLE 5.6

Quadratus Lumborum Muscle

Muscle	Origin	Insertion	Action	Innervation
Quadratus lumborum	Medial part of iliac crest	12th rib, lower lumbar vertebrae	Lateral flexion of vertebral column; fixes last rib to form stable base for contraction of diaphragm	Branches from T12 and L1-L3

TABLE 5.7

Hip Extensor Muscles

Muscle	Origin	Insertion	Action	Innervation
Gluteus maximus	Lateral surface of ilium behind posterior gluteal line, dorsal sacroiliac and sacrotuberous ligaments, dorsal surface of sacrum	Iliotibial tract, gluteal tuberosity of femur	Extension, lateral rotation, abduction (upper fibers), and adduction (lower fibers) of thigh	Inferior gluteal nerves L5-S2
Semi-tendinosus	Ischial tuberosity	Medial surface of proximal end of tibia	Extension of thigh, flexion of leg, medial rotation of flexed leg	Sciatic nerve: tibial branch L4-S2
Semi-membranosus	Ischial tuberosity	Medial condyle of tibia	Extension of thigh, flexion of leg, medial rotation of flexed leg	Sciatic nerve: tibial branch L4-S2
Biceps femoris (long head)	Long head: ischial tuberosity	Lateral head of fibula and lateral tibial condyle	Extension of thigh (long head), flexion of leg, lateral rotation of flexed leg	Sciatic nerve: tibial branch to long head L5-S3
Secondary muscles	Adductor magnus (inferior portion) and gluteus medius (posterior portion)			

TABLE 5.8

Hip Flexor Muscles

Muscle	Origin	Insertion	Action	Innervation
Iliacus (primary)	Iliac fossa	Lesser trochanter (with psoas major) of femur	Flexion of thigh, slight adduction of the thigh of free limb	Femoral nerve L1-L4
Psoas (primary)	Bodies and transverse processes of lumbar vertebrae	Lesser trochanter of femur	Flexion of thigh, slight adduction of the thigh of free limb	L1-L4
Rectus femoris	Anterior inferior iliac spine, ilium, and above the acetabulum	Patella and through patellar ligament to tibial tuberosity	Extension of leg, flexion of thigh	Femoral nerve L1-L4
Tensor fasciae latae	Iliac crest posterior to anterior superior iliac spine (ASIS)	Iliotibial tract	Flexion, medial rotation, and abduction of thigh	Superior gluteal nerve L4-S1
Sartorius	ASIS	Medial surface, proximal end of tibia just distal to tibial tuberosity	Flexion, abduction, and lateral rotation of thigh; flexion of leg	Femoral nerve
Pectineus	Superior ramus of pubis	Femur just distal to lesser trochanter	Flexion and adduction of thigh	Femoral nerve, possibly obturator or accessory obturator nerve or both
Adductor brevis	Body and inferior ramus of pubis	Pectinial line, proximal part of linea aspera of femur	Adduction and flexion of thigh	Obturator nerve
Adductor longus	Pubic tubercle	Medial lip of linea aspera of femur	Adduction and flexion of thigh	Obturator nerve
Adductor magnus (superior fibers)	Inferior ramus of pubis, ramus of ischium, ischial tuberosity	Linea aspera (anterior fibers), adductor tubercle of femur (posterior fibers)	Adduction, flexion (anterior fibers), and extension (posterior fibers) of thigh	Obturator nerve (anterior fibers), sciatic nerve (posterior fibers)
Gluteus medius (anterior)	Lateral surface of ilium between anterior and posterior gluteal line	Greater trochanter	Abduction, medial rotation and flexion (anterior fibers), and lateral rotation and extension (posterior fibers) of thigh	Superior gluteal nerves L5-S1
Piriformis	Sacrum (pelvic surface)	Greater trochanter of femur	Lateral rotation of thigh of extended hip, abduction of thigh when thigh is flexed	Ventral rami S1 and S2

TABLE 5.9

Hip External Rotator Muscles

Muscle	Origin	Insertion	Action	Innervation
Obturator externus	Obturator membrane, bone around obturator foramen on external surface of pelvis	Trochanteric fossa of femur	Lateral rotation of thigh	Obturator nerve
Obturator internus	Obturator membrane, bone around obturator foramen on internal surface of pelvis	Medial surface of greater trochanter above trochanteric fossa of femur	Lateral rotation of thigh, abduction of thigh when thigh is flexed	Nerve to obturator internus
Quadratus femoris	Ischial tuberosity	Posterior surface of femur between greater and lesser trochanters	Lateral rotation and adduction of thigh	Nerves to quadratus femoris
Piriformis	Sacrum (pelvic surface)	Greater trochanter of femur	Lateral rotation of thigh of extended hip, abduction of thigh when thigh is flexed	Ventral rami S1 and S2
Superior gemellus	Ischial spine	Superior borders of obturator internus tendon	Lateral rotation of thigh, abduction of thigh when thigh is flexed	Nerve to obturator internus
Inferior gemellus	Ischial tuberosity	Inferior border of obturator internus tendon	Lateral rotation of thigh, abduction of thigh when thigh is flexed	Nerve to quadratus femoris
Gluteus maximus	Lateral surface of ilium behind posterior gluteal line, dorsal sacroiliac and sacrotuberous ligaments, dorsal surface of sacrum	Iliotibial tract, gluteal tuberosity of femur	Extension, lateral rotation, abduction (upper fibers), and adduction (lower fibers) of thigh	Inferior gluteal nerves
Secondary muscles	Sartorius, biceps femoris (long head), gluteus medius (posterior), psoas major, adductor magnus (position dependent), adductor longus, popliteus (with tibia fixed)			

TABLE 5.10

Hip Internal Rotator Muscles

Muscle	Origin	Insertion	Action	Innervation
Gluteus minimus (anterior fibers)	Lateral surface of ilium between anterior and inferior gluteal lines	Greater trochanter of femur	Abduction, medial rotation, and flexion of thigh	Superior gluteal nerves L4-S1
Tensor fasciae latae	Iliac crest posterior to ASIS	Iliotibial tract	Flexion, medial rotation, and abduction of thigh	Superior gluteal nerves L4-S1
Gluteus medius (anterior fibers)	Lateral surface of ilium between anterior and posterior gluteal line	Greater trochanter	Abduction, medial rotation and flexion (anterior fibers), and lateral rotation and extension (posterior fibers) of thigh	Superior gluteal nerves L5-S1
Secondary muscles	Semitendinosus, semimembranosus, adductor magnus (position dependent), adductor longus (position dependent)			

TABLE 5.11

Hip Adductor Muscles

Muscle	Origin	Insertion	Action	Innervation
Pectineus	Superior ramus of pubis	Femur just distal to lesser trochanter	Flexion and adduction of thigh	Femoral nerve L2-L3
Adductor longus	Pubic tubercle	Medial lip of linea aspera of femur	Adduction and flexion of thigh	Obturator nerve L2-4
Gracilis	Inferior ramus of pubis, ramus of ischium	Medial surface, proximal end of tibia just distal to medial condyle	Adduction of thigh, flexion of leg, medial rotation of flexed leg	Obturator nerve L3-L4
Adductor brevis	Body and inferior ramus of pubis	Pectinial line, proximal part of linea aspera of femur	Adduction and flexion of thigh	Obturator nerve L2-4
Adductor magnus	Inferior ramus of pubis, ramus of ischium, ischial tuberosity	Linea aspera (anterior fibers), adductor tubercle of femur (posterior fibers)	Adduction, flexion (anterior fibers), and extension (posterior fibers) of thigh	Obturator nerve (anterior fibers), sciatic nerve (posterior fibers) L2-4

TABLE 5.12

Hamstring Muscles

Muscle	Origin	Insertion	Action	Innervation
Semitendinosus	Ischial tuberosity	Medial surface of proximal end of tibia	Extension of thigh, flexion of leg, medial rotation of flexed leg	Sciatic nerve: tibial branch L4-S2
Semimembranosus	Ischial tuberosity	Medial condyle of tibia	Extension of thigh, flexion of leg, medial rotation of flexed leg	Sciatic nerve: tibial branch L4-S2
Biceps femoris	Long head: ischial tuberosity Short head: linea aspera of femur and lateral intermuscular septum	Lateral head of fibula and lateral tibial condyle	Extension of thigh (long head), flexion of leg, lateral rotation of flexed leg	Sciatic nerve: tibial branch to long head (L5-S3), common fibular branch to short head (L5-S2)

TABLE 5.13

Intratester Reliability of the Active Hamstring 90/90 Test

Study	n	Sample	r*	ICC**
Gajdosik & Lusin 1983	15	Healthy males (18-26 years)	0.99R 0.99L	NT NT
Sullivan et al. 1992	12	Healthy adults	NT	0.99
Worrell et al. 1994	10	Healthy adults	NT	0.93
Webright et al. 1997	12	Healthy adults	NT	0.98
Gajdosik et al.1993	30	Healthy adults (18-40 years)	NT	0.86

NT = Not tested.

*Pearson's r.

**Intraclass correlation.

Data taken from: N.B. Reese and W.D. Bandy (Eds), 2002, *Joint range of motion and muscle length testing* (Philadelphia, PA, W.B. Saunders Co.).

TABLE 5.14

Intratester Reliability of the Straight Leg Raise Test

Study	n	Sample	r*	ICC**
Hsieh et al. 1983	10	Healthy adults (26-30 years)	0.95	NT
Rose 1991	18	Healthy adults (x = 19.5 years)	0.86R 0.83L	NT NT
Wang et al. 1993	10	Healthy adults (18-37 years)	NT NT	0.90 dominant 0.91 nondominant
Hanton & Chandler 1994	75	Healthy females (18-29 years)	NT	0.91
Gajdosik et al. 1993	30	Healthy adults (18-40 years)	NT	0.83

NT = Not tested.

*Pearson's r.

**Intraclass correlation.

Data taken from: N.B. Reese and W.D. Bandy (Eds), 2002, *Joint range of motion and muscle length testing* (Philadelphia, PA, W.B. Saunders Co.).

TABLE 5.15

Intratester Reliability of the Passive Hamstring 90/90 Test

Study	n	Sample	r*	ICC**
Bandy & Irion 1994	15	Healthy adults (22-36 years)	0.91	NT
Bandy et al. 1997	20	Healthy adults (20-40 years)	0.97	NT
Gajdosik et al. 1993	30	Healthy adults (18-40 years)	NT	0.90

NT = Not tested.

*Pearson's r.

**Intraclass correlation.

Data taken from: N.B. Reese and W.D. Bandy (Eds), 2002, *Joint range of motion and muscle length testing* (Philadelphia, PA, W.B. Saunders Co.).

TABLE 5.16

Gastrosoleus Muscle Group

Muscle	Origin	Insertion	Action	Innervation
Gastrocnemius	Posterior surface of femur (medial and lateral condyles), capsule of knee joint	Calcaneus via Achilles tendon (tendo calcaneus), posterior calcaneus	Plantarflexion of foot, flexion of knee	Tibial nerve S1-S2
Soleus	Fibula (head, posterior aspect, and proximal third of shaft), tibia (popliteal line), aponeurosis between tibia and fibula	Calcaneus via Achilles tendon (tendo calcaneus), posterior calcaneus	Plantarflexion of foot	Tibial nerve S1-S2

Fundamental Movement Testing

All movement of the human body consists of fundamental and varying degrees of complex movements. Fundamental movements are essential to successful completion of more complex and multiplanar movements required in sport, occupations, and exercise. Synergistic movements occur throughout the body, and these movements are only as strong as the "weakest link" in the respective kinetic chain. As we have previously mentioned, it is difficult to objectively assess "function." Assessment of components of function is usually easier and more objective.

Two main components of fundamental movement analysis are addressed in this chapter: movement analysis and the Functional Movement Screen (FMS). It is suggested that the clinician use these assessments in the order of their listing (specifically the movement analyses prior to the FMS), as they progress in difficulty. If a client is unsuccessful with less complex movement skills, he or she will also most likely be unsuccessful with the more complex skills listed later.

Movement Analysis

The movement analyses are described in terms of patterns of muscle activation. These assessments look at specific muscle synergistic actions and the relationships between various muscle groups, such as agonistic, antagonistic, and synergistic muscle groups, and the specified body movements. With the specific movement analyses described in this chapter there is an ideal pattern of muscle activation that is termed "normal." These muscle firing patterns and coordination between the different muscle groups are expected of normal movement synergy. Altered muscle firing patterns are theorized to promote dysfunctional movement patterns for the fundamental movement being assessed. In outlining each movement assessment we list the required order of muscle recruitment for a normal score, as well as the altered muscle firing patterns that the clinician can expect to encounter.

CERVICAL DEEP FLEXOR MUSCLE ASSESSMENT

▶ **Purpose:** To assess muscle balance among various cervical flexion muscles. For the origin, insertion, and innervation of these muscles, see table 6.1 on page 100.

▶ **Equipment:** Solid, stable testing surface.

Procedure (Jull & Janda 1987)

1. Position the client so that he or she is supine with the head on the table without a pillow.
2. Instruct the client to lift the head and to look down at his or her feet.
3. Observe the flexion.

Analysis and Interpretation of Data

■ A normal gradual segmental flexion should occur along the length of the cervical spine.

■ Substitution by scalene and sternocleidomastoid muscles is noted when the client lifts the head straight up toward the ceiling initially (upper cervical extension with lower cervical flexion).

Statistics

There are no known reliability data for this assessment.

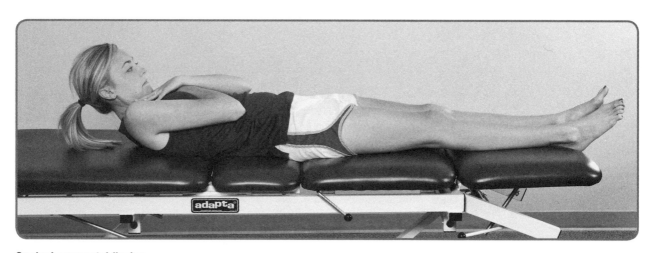

Gradual segmental flexion.

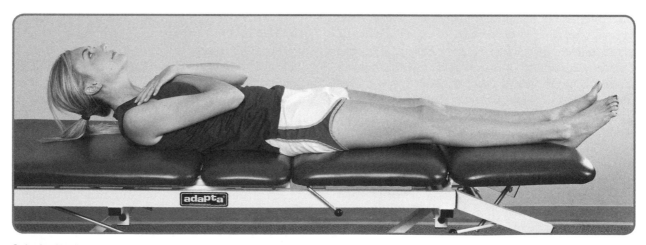

Substitution by scalene and sternocleidomastoid muscles.

HIP EXTENSION ASSESSMENT

▶ **Purpose:** To assess for appropriate hip extension muscle firing patterns. For the origin, insertion, and innervation of these muscles, see table 6.2 on pages 100-101.

▶ **Equipment:** Solid, stable testing surface.

Procedure (Clark 2001)

1. Position the client so that he or she is prone with legs extended and arms relaxed.
2. Stand at the side of the test leg.
3. With the thumb and index finger of one hand, palpate bilateral erector spinae while palpating the muscle belly of the gluteus maximus and hamstring with the index finger and thumb of the opposite hand.
4. Instruct the client to extend the hip off the table while keeping the knee straight.

Analysis and Interpretation of Data

Normal firing order:

1. Gluteus maximus
2. Opposite erector spinae
3. Same erector spinae and hamstring

Altered firing pattern:

1. Weak agonist = gluteus maximus
2. Overactive antagonist = psoas
3. Overactive stabilizer = erector spinae
4. Overactive synergist = hamstring

Statistics

There are no known reliability data for this assessment.

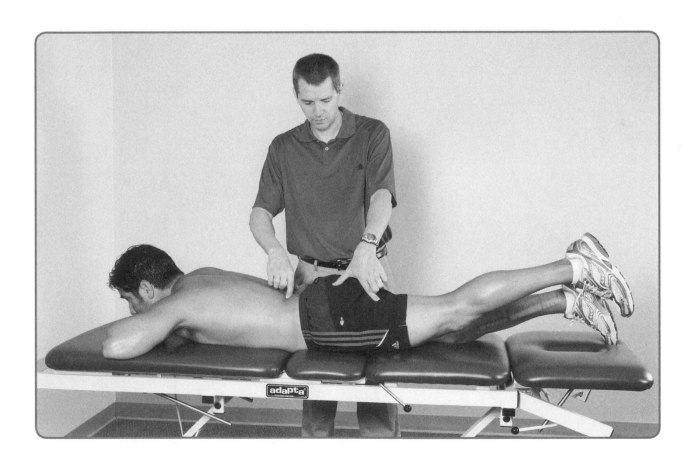

HIP ABDUCTION ASSESSMENT

▶ **Purpose:** To assess for appropriate hip abduction muscle firing patterns. For the origin, insertion, and innervation of these muscles, see table 6.3 on page 101.

▶ **Equipment:** Solid, stable testing surface.

Procedure

1. Position the client so that he or she is side-lying with bottom leg hip and knee flexed slightly for stability.
2. Stand at side of the test leg.
3. Palpate the quadratus lumborum with one hand and the tensor fasciae latae (TFL) and gluteus medius with the other hand.
4. Instruct the client to abduct the leg while keeping the knee straight.
5. Monitor for movement out of the frontal plane.

Analysis and Interpretation of Data

Normal firing order:

1. Gluteus medius
2. TFL–quadratus lumborum

Altered firing pattern:

1. Weak agonist = gluteus medius
2. Overactive antagonist = adductors
3. Overactive synergist = TFL
4. Overactive stabilizer = quadratus lumborum
5. Overactive neutralizer = piriformis

Statistics

There are no known reliability data for this assessment.

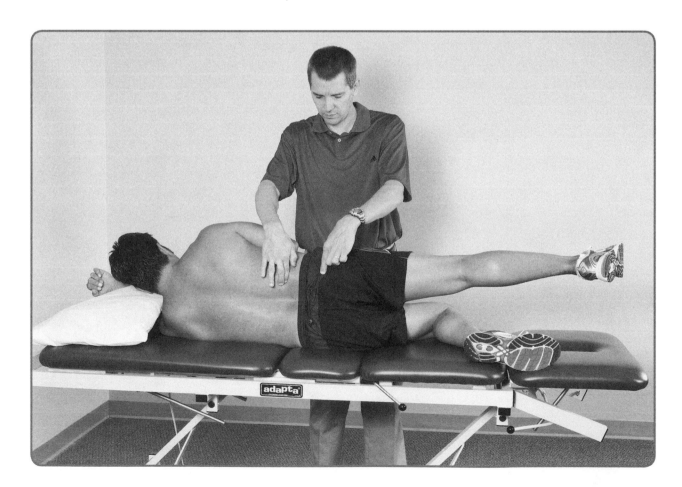

TRUNK FLEXION ASSESSMENT

▶ **Purpose:** To assess for appropriate hip flexion muscle firing patterns. For the origin, insertion, and innervation of these muscles, see table 6.4 on page 102.

▶ **Equipment:** Solid, stable testing surface.

Procedure (Clark 2001)

1. Position the client so that he or she is supine with bilateral knees and hips flexed to 90°.
2. Stand at the client's side, directly facing the client.
3. Palpate the abdominal muscle group with the closest hand.
4. Instruct the client to perform a normal curl-up (i.e., to flex segmentally from head to waist).

Analysis and Interpretation of Data

Normal firing pattern:

1. The client should be able to maintain a drawing-in maneuver while performing a curl-up.
2. The inability to maintain the drawing-in position demonstrates an altered firing pattern of the abdominal stabilization mechanism.

Altered firing pattern:

1. Weak agonist = abdominal complex
2. Overactive antagonist = erector spinae
3. Overactive synergist = psoas

Statistics

There are no known reliability data for this assessment.

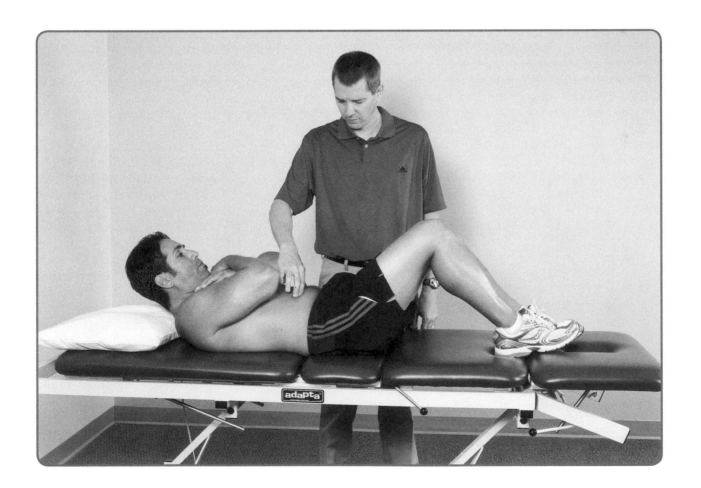

Functional Movement Screen

With contributions from Gray Cook, MSPT, OCS, CSCS; Lyle Burton, PhD, ATC, CSCS; and Kyle Kiesel, PT, PhD, ATC, CSCS

The FMS (Cook et al. 1998) is designed to identify individuals who have developed compensatory movement patterns in the kinetic chain. It was initially intended to screen for movement compensations among nonsymptomatic individuals to serve as a basis for the proper design of an exercise intervention specific to individual needs. The seven movements in the FMS are an effort to challenge the body's ability to facilitate movement through the proximal-to-distal sequence. Once an inefficient movement pattern has been isolated by the FMS, functional prevention strategies can be instituted to avoid problems such as imbalance, micro-traumatic breakdown, and injury (Cook et al. 1998). The screen can be implemented with a large target audience in a fairly efficient manner. It does require careful observation by a trained clinician and should be critically scored. The authors of this book have also used it for evaluation of movement patterns during the rehabilitation process as a screening tool for movement dysfunction. Clinicians who determine a movement dysfunction during one of these assessments can pursue additional testing to discern the exact cause of the dysfunctional movement pattern. In this way the FMS can be used as a general assessment, allowing clinicians to use further testing at their disposal to diagnose movement dysfunction.

The scoring of the FMS consists of four possibilities. The scores range from zero to III, with III being the best possible score.

■ A score of zero is given if at any time during the testing the client has pain anywhere in the body.

■ A score of I is given if the person is unable to complete the movement pattern or is unable to assume the position to perform the movement.

■ A score of II is given if the person is able to complete the movement but must compensate in some way to perform the fundamental movement.

■ A score of III is given if the person performs the movement correctly without compensation.

The scoring sheet contains an area to be used for comments; clinicians should utilize this area when scoring to make notes about the client's specific movement problems (Cook et al. 1998). The developers of the FMS suggest that if in doubt about scoring between two levels, one should score the lower level and be critical in one's assessments.

Three tests with additional clearing screens are performed; these are graded as positive or negative. These screens are general assessments done to ensure safety of further testing. These clearing movements take into account only pain (positive if the client has pain, negative if the client does not have pain). If a client has a positive clearing screen test, then the score will be zero.

The reliability of the FMS has been examined in two separate studies. The first study used multiple raters who viewed videos of subjects performing each of the seven tests. Each test was analyzed individually for its reliability, and the composite score (out of the possible 21) was analyzed as a continuous variable. The results for the composite score were excellent, with an interclass correlation coefficient of 0.98. Analyzing the composite score of the FMS on a continuous scale is acceptable when one is dealing with group data as in an injury prediction study (Kiesel 2006); but because the FMS is scored on an ordinal scale, the appropriate analysis for assessing its reliability is with the kappa statistic (Portney & Watkins 2000). The kappa statistic is a chance-corrected level of agreement that compares how sets of raters score the subjects on each test and factors out chance agreement. The kappa coefficients are scaled differently than the intraclass correlation coefficients (ICCs), and it is important that the reader understand this difference when interpreting reliability studies. For example, a value of 0.60 is considered quite low for an ICC, while a kappa of 0.60 represents the lower end of substantial agreement.

The second study (Minick et al. 2007) assessed the reliability of the FMS using the kappa statistic. The authors had two pairs of raters score each of the seven tests of the FMS on 39 subjects. The results for the tests ranged from a kappa of 0.75 to 1.0 (substantial to perfect agreement). The lowest value was on the lunge test, while the squat and shoulder mobility tests showed perfect agreement of 1.0. The overall average kappa value for the seven tests between raters was 0.90, which represents almost perfect agreement (Sim & Wright 2005).

The seven movements in the FMS are the deep squat, the hurdle step, the in-line lunge, shoulder mobility, active straight leg raise, trunk stability push-up, and rotational stability. A scoring sheet for the FMS can be found on page 276 in the appendix; you may reproduce the scoring sheet for your own use. (For your convenience, the scoring sheet is also included on the DVD that accompanies this book.)

DEEP SQUAT ASSESSMENT

▶ **Purpose:** To assess bilateral symmetrical mobility of the hips, knees, and ankles. The dowel held overhead is used to assess bilateral symmetrical mobility of the shoulders as well as the thoracic spine. The deep squat is a test that challenges total-body mechanics when performed properly.

▶ **Equipment:** Dowel rod, cane, or straight-line stick.

Procedure

1. Position the client so that he or she is standing with feet approximately shoulder-width apart and holding the dowel so that a 90° angle is formed at bilateral elbows when the dowel is just above the head.

2. Have the client press the dowel overhead until bilateral elbows are extended.

3. Instruct the client to gradually descend into a squat position with bilateral heels on the floor and the dowel extended maximally overhead.

4. Have the client perform as many as three repetitions.

5. If the criterion for a score of II is not achieved, ask the client to perform the test with a 2 × 6 under the heels.

Analysis and Interpretation of Data

Repetitive squatting for clients with and without low back pain as a test of endurance (maximum number of repetitions to a maximum of 50) revealed that physical capacity was reduced with advancing age (Alaranta et al. 1994).

Grade	Grading criteria (Cook et al. 1998)
III	• Upper torso parallel with tibia or toward vertical • Femur below horizontal • Knees aligned over feet • Dowel aligned over feet
II	• 2 × 6 under heels • Upper torso parallel with tibia or toward vertical • Femur below horizontal • Knees aligned over feet • Dowel aligned over feet
I	• 2 × 6 under heels • Tibia and upper torso not parallel • Femur not below horizontal • Knees not aligned over feet • Lumbar flexion noted
0	• Given if pain with any portion of testing

Compensation.

Limited mobility in upper torso.

(continued)

(continued from previous page)

Statistics

■ Repetitive squatting with feet 15 cm apart, to the point that thighs were parallel to the floor, performed at a rate of one repetition per 2 to 3 s (with a maximum number of repetitions at 50): $r = 0.95$ ($n = 32$) (Alaranta et al. 1994).

■ Refer to the introduction to this section of the chapter for reliability data.

Notes

■ Poor performance of this test can be the result of several factors.

■ Limited mobility in the upper torso can be attributed to poor glenohumeral or thoracic spine mobility or both.

■ Limited mobility in the lower extremity, including poor closed kinetic chain dorsiflexion of the ankles or poor flexion of the hips, may also cause poor test performance.

■ Testing by FMS developers suggests that when a client achieves a score of II, minor limitations exist either with closed kinetic chain dorsiflexion of the ankle or extension of the thoracic spine.

■ When a client achieves a score of I or less, gross limitations may exist with the motions previously mentioned as well as hip flexion.

■ For specific suggestions on exercise corrections or additional information, please consult Cook (1998) or visit the Web site www.functionalmovement. com.

HURDLE STEP ASSESSMENT

▶ **Purpose:** To assess bilateral mobility and stability of the hips, knees, and ankles and proper stride during a stepping motion. The movement requires proper coordination and stability between the hips and torso during the stepping motion as well as single-leg stance stability (Cook et al. 1998).

▶ **Equipment:** Dowel rod, cane, or straight-line stick; PVC pipe hurdle or colored tape or string.

Procedure

1. Position the client so that he or she is standing with bilateral feet approximately shoulder-width apart.
2. Adjust the hurdle to the height of the client's tibial tuberosity (alternatively, colored tape or string in a door frame can be used at the same height).
3. Align the client's toes directly beneath the hurdle (or tape or string).
4. Ask the client to step over the barrier and touch the heel of the step leg to the floor while keeping the stance knee and hip extended.
5. Perform the assessment as many as three times bilaterally.

Analysis and Interpretation of Data

If one repetition is completed bilaterally and meets the following criteria, the corresponding score is given.

Statistics

Refer to the introduction to this section for reliability data.

Grade	Grading criteria (Cook et al. 1998)
III	• Hips, knees, and ankles remain aligned in the sagittal plane • Movement in lumbar spine is minimal • Dowel and hurdle remain parallel
II	• Alignment is lost between hips, knees, and ankles • Movement in lumbar spine occurs • Dowel and hurdle do not remain parallel
I	• Foot contacts hurdle • Loss of balance occurs at any time
0	• This score is given if pain is noted with any portion of testing

Beginning.

Stepping over barrier.

Compensation.

(continued)

(continued from previous page)

Notes

- Poor performance during this testing can be the result of several factors. It may simply be due to poor stability of the stance leg or poor mobility of the step leg.
- Imposing maximal hip flexion of one leg while maintaining apparent hip extension of the opposite leg requires the client to demonstrate relative bilateral asymmetric hip mobility (Cook et al. 1998).

- According to testing by FMS developers, a score of II suggests that limitations most often exist with ankle dorsiflexion or hip flexion with the step leg or both.
- With a score of I or less, relative asymmetric hip immobility may exist secondary to an anteriorly tilted pelvis.
- For specific suggestions on exercise corrections or additional information, please consult Cook and colleagues (1998) or visit the Web site www.functionalmovement.com.

IN-LINE LUNGE ASSESSMENT

▶ **Purpose:** This test is aimed at placing the body in a position that will focus on the stresses incurred during rotational, decelerating, and lateral-type movements. The test is used to assess hip mobility and stability, as well as ankle and knee stability.

▶ **Equipment:** Dowel rod, cane, or straight-line stick; 2 × 6; tape measure.

Procedure

1. Measure the client's tibial length with a tape measure.

2. Have the client place one foot on the end of the 2 × 6.

3. Mark the distance of the client's tibial length on the 2 × 6 from the client's back toe.

4. Place the dowel rod behind the client's back, touching the head, thoracic spine, and sacrum.

5. Instruct the client to grasp the top of the dowel with the hand ipsilateral to the back foot.

6. Ask the client to take a step and place the front heel on the mark.

7. Ensure that the client's feet are on the same line and pointing straight ahead throughout the assessment.

8. Have the client perform the lunge up to three times in a slow, controlled manner.

9. Also assess with bilateral lower extremities in alternate positions.

Analysis and Interpretation of Data

If one repetition is completed that meets the following criteria, the corresponding score is given.

Grade	Grading criteria (Cook et al. 1998)
III	• Minimal to no torso movement occurs • Feet remain in sagittal plane on the 2 × 6 • Knee touches behind heel of front foot
II	• Movement is noted in torso • Feet do not remain in sagittal plane • Knee does not touch behind heel of front foot
I	• Loss of balance occurs at any time
0	• This score is given if pain is associated with any portion of the test

Statistics

Refer to the introduction to this section for reliability data.

Notes

■ Poor performance during this test can be the result of several factors. Hip mobility may be inadequate in either the stance or the step leg; the stance-leg knee or ankle may not have the required stability as the client performs the lunge; or an imbalance between relative adductor weakness and abductor tightness in one or both hips may cause poor test performance.

■ Limitations in the thoracic spine region may inhibit the client from performing the test properly (Cook et al. 1998).

■ Testing by FMS developers suggests that a score of II reveals minor limitations with adduction of one or both hips.

■ A score of I or less indicates that a relative adductor weakness and abductor tightness may be present around one or both hips.

■ For specific suggestions on exercise corrections or additional information, please consult Cook and colleagues (1998) or visit the Web site www.functionalmovement.com.

SHOULDER MOBILITY ASSESSMENT

▶ **Purpose:** To assess bilateral shoulder range of motion combining internal rotation with adduction and external rotation with abduction.

▶ **Equipment:** Dowel rod, cane, or straight-line stick; tape measure.

Procedure

1. Measure the client's hand length from the wrist crease to the tip of the third digit.
2. Instruct the client to make a fist in bilateral hands with thumbs inside the fists.
3. Instruct the client to reach with the upper extremities behind the back in an attempt to touch hands together.
4. Measure the distance between the two fists.
5. Perform the assessment as many as three times bilaterally.

Analysis and Interpretation of Data

Grade	Grading criteria (Cook et al. 1998)
III	• Fists are within one hand length.
II	• Fists are within 1½ hand lengths.
I	• Fists fall more than 1½ lengths.
0	• This score is given if pain is associated with any portion of the test.

Statistics

Refer to the introduction to this section for reliability data.

Notes

- A shoulder stability screen (clearing exam) should be performed even if the client scores a III.
- An active impingement test should also be performed.
- Poor performance during this testing can have several causes. One widely accepted explanation is that increased external rotation is gained at the expense of internal rotation in overhead throwing athletes. Excessive development and shortening of the pectoralis minor or latissimus dorsi muscles (see chapter 5) can cause postural alterations of forward or rounded shoulders. Finally, a scapulothoracic dysfunction may be present, resulting in decreased glenohumeral mobility secondary to poor scapulothoracic mobility or stability (Cook et al. 1998).
- Testing by FMS developers has identified that minor postural changes or shortening of isolated axiohumeral or scapulohumeral muscles exist with a score of II.
- When a client scores a I or zero, a scapulothoracic dysfunction may exist.
- For specific suggestions on exercise corrections or additional information, please consult Cook and colleagues (1998) or visit the Web site www.functionalmovement.com.

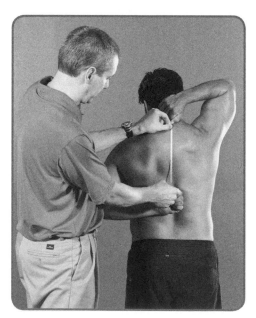
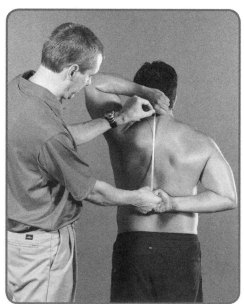

ACTIVE STRAIGHT LEG RAISE ASSESSMENT

▶ **Purpose:** To assess active hamstring and gastrocsoleus flexibility while the individual maintains a stable pelvis.

▶ **Equipment:** Dowel rod, cane, or straight-line stick; 2 × 6.

Procedure

1. Position the client so that he or she is supine with arms at side, palms up, and head flat on floor.
2. Place a 2 × 6 under the client's knees.
3. Instruct the client to lift the test leg with a dorsiflexed ankle and an extended knee.
4. Ensure that the opposite knee remains in contact with the 2 × 6 and that the head and shoulders remain flat on the floor.
5. At the client's end-range position, align a dowel along the medial malleolus of the test leg and perpendicular to the floor.
6. Perform the assessment as many as three times bilaterally.

Analysis and Interpretation of Data

Grade	Grading criteria (Cook et al. 1998)
III	• Dowel dropped down from medial malleolus resides between midthigh and anterior superior iliac spine.
II	• Dowel dropped down from medial malleolus resides between midthigh and midpatella.
I	• Dowel dropped down from medial malleolus resides below midpatella.
0	• This score is given if pain is noted with any portion of testing.

Statistics

Refer to the introduction to this section for reliability data.

Notes

■ Poor performance during this test can be the result of several factors: The client may have poor functional hamstring flexibility (see chapter 5) or may have inadequate mobility of the opposite hip, stemming from iliopsoas inflexibility associated with an anteriorly tilted pelvis (Cook et al. 1998).

■ According to the developers of the FMS, a score of II suggests that minor asymmetric hip mobility limitations or moderate isolated, unilateral muscle tightness may exist.

■ When a client scores a I or zero, relative hip mobility limitations are gross.

■ Passive and dynamic hamstring flexibility exercises and warm-up drills are suggested for clients with a score below III.

■ For specific suggestions on exercise corrections or additional information, please consult Cook and colleagues (1998) or visit the Web site www.functionalmovement.com.

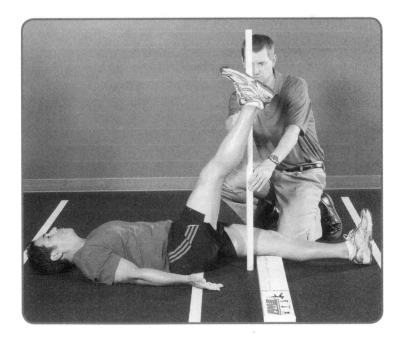

TRUNK STABILITY PUSH-UP ASSESSMENT

▶ **Purpose:** To assess the ability to stabilize the spine in an anterior and posterior plane during a closed-chain upper body movement. This test is used to assess trunk stability in the sagittal plane while a symmetrical upper extremity motion is performed.

▶ **Equipment:** No specific equipment required.

Procedure

1. Position the client so that he or she is prone on the floor with hands placed shoulder-width apart at the appropriate position (see table), with knees fully extended.
2. Instruct the client to perform one push-up in this starting position.
3. Instruct the client to lift the body as a unit, with no "lag" in the lumbar spine.
4. If the client cannot perform a push-up in this position, have him or her lower the hands to the appropriate position (see table) and perform a reassessment.
5. You can perform the assessment as many as three times.

Analysis and Interpretation of Data

Grade	Grading criteria (Cook et al. 1998)
III	• Males perform one repetition with thumbs above head. • Females perform one repetition with thumbs in line with chin.
II	• Males perform one repetition with thumbs in line with chin. • Females perform one repetition with thumbs in line with clavicle.
I	• Males are unable to perform one repetition with hands in line with chin. • Females are unable to perform one repetition with thumbs in line with clavicle.
0	• This score is given if pain is noted with any portion of testing.

Statistics

Refer to the introduction to this section for reliability data.

Notes

■ Lumbar extension should also be cleared after this test, even with a grade of III. This can be done with a prone press-up.

■ Poor performance during this test can be attributed simply to poor stability of the trunk stabilizers (Cook et al. 1998).

■ According to testing by FMS developers, a score of II suggests that mild to moderate limitations exist with symmetric trunk stability.

■ When a client scores a I or less, severe limitations exist with regard to symmetric trunk stability.

■ For specific suggestions on exercise corrections or additional information, please consult Cook and colleagues (1998) or visit the Web site www.functionalmovement.com.

ROTATIONAL STABILITY ASSESSMENT

▶ **Purpose:** To assess multiplanar trunk stability while a combined upper and lower extremity motion is performed.

▶ **Equipment:** 2 × 6.

Procedure

1. Position the client in a quadruped position with shoulders and hips at 90° relative to the upper torso. Bilateral knees are positioned at 90° and ankles remain dorsiflexed.

2. Place the 2 × 6 between the knees and hands so that they are in contact with the board.

3. Instruct the client to flex the shoulder and extend the same-side hip and knee (first position).

4. Have the client raise the leg and hand enough to clear the testing surface by approximately 6 in. (15 cm).

5. Ensure that the lifted extremities remain in the same plane as the 2 × 6.

6. Instruct the client to flex the same shoulder and knee so that they touch (second position).

7. Perform the assessment bilaterally for up to three repetitions.

8. If a III is not attained, have the client perform a diagonal pattern using the opposite shoulder and hip in the same manner as already described.

Analysis and Interpretation of Data

Grade	Grading criteria (Cook et al. 1998)
III	• Client performs one unilateral repetition while keeping torso parallel to board. • Knee and elbow touch in line with board.
II	• Client performs one diagonal repetition while keeping torso parallel to board. • Knee and elbow touch in line with board.
I	• Client is unable to perform diagonal repetitions.
0	• This score is given if client is unable to attain the correct position and if pain is associated with any portion of the testing.

Statistics

Refer to the introduction to this section for reliability data.

First position.

Second position.

Notes

▪ Poor performance during this testing can be attributed simply to poor asymmetric stability of the trunk stabilizers (Cook et al. 1998).

▪ According to testing by FMS developers, a score of II indicates mild to moderate limitations with asymmetric trunk stability.

▪ When a client scores I or less, severe limitations exist with regard to asymmetric trunk stability.

▪ Lumbar flexion should be cleared after this test, even with a score of III. This can be accomplished with a hands-and-knees (quadruped) position and rocking pelvis to heels.

▪ For specific suggestions on exercise corrections or additional information, please consult Cook and colleagues (1998) or visit the Web site www.functionalmovement.com.

TABLE 6.1

Capital and Cervical Neck Flexor Muscles

Muscle	Origin	Insertion	Action	Innervation
CAPITAL FLEXION				
Rectus capitis anterior	Lateral mass of atlas	Occipital bone	Stabilization of occipital-atlantal joint, flexion of head	Ventral rami of C1 and C2
Rectus capitis lateralis	Transverse process of C1	Occipital bone	Stabilization of occipital-atlantal joint, lateral flexion of head	Ventral rami of C1 and C2
Longus capitis	Transverse processes of C3-C6	Occipital bone	Flexion of head and upper cervical vertebrae	Ventral rami of C1-C3
Secondary muscles	Suprahyoid muscles: mylohyoid, stylohyoid, geniohyoid, digastric			
CERVICAL FLEXION				
Sternocleidomastoid	Mastoid process and lateral superior nuchal line	Sternal head: anterior manubrium Clavicular head: superior medial clavicle	Neck flexion, ipsilateral side-bend, contralateral rotation	Spinal root of accessory nerve C2-C3
Longus coli	Bodies of T1-T3, transverse processes of C3-C5, bodies of C5-T3	Transverse processes of C5 and C6, anterior surface of C1, bodies of C2-C4 (respectively with listed origins)	Flexion, possibly lateral flexion of neck	Ventral rami of C2-C6
Anterior scalene	Anterior tubercles of transverse processes of C4-C6	1st rib (scalene tubercle)	Elevation of 1st rib and ipsilateral side-bend, contralateral rotation	C4-C6
Secondary muscles	Middle scalene, posterior scalene Infrahyoids: sternothyroid, thyrohyoid, sternohyoid, omohyoid			

TABLE 6.2

Muscles for Hip Extension Analysis

Muscle	Origin	Insertion	Action	Innervation
Gluteus maximus	Lateral surface of ilium behind posterior gluteal line, dorsal sacroiliac and sacrotuberous ligaments, dorsal surface of sacrum	Iliotibial tract, gluteal tuberosity of femur	Extension, lateral rotation, and abduction (upper fibers) and adduction (lower fibers) of thigh	Inferior gluteal nerves L5-S2
Erector spinae Iliocostalis lumborum	Iliac crest, sacrum	Lower borders of lowest seven ribs	Bilateral action: extension of trunk Unilateral action: lateral trunk flexion ipsilaterally	Dorsal rami of spinal nerves in corresponding area
Longissimus thoracis	Intermediate part of extensor aponeurosis	Lower 10 ribs and adjacent vertebral transverse processes		
Semitendinosus	Ischial tuberosity	Medial surface of proximal end of tibia	Extension of thigh, flexion of leg, medial rotation of flexed leg	Sciatic nerve: tibial branch L4-S2

Muscle	Origin	Insertion	Action	Innervation
Semimembranosus	Ischial tuberosity	Medial condyle of tibia	Extension of thigh, flexion of leg, medial rotation of flexed leg	Sciatic nerve: tibial branch L4-S2
Biceps femoris	Long head: ischial tuberosity Short head: linea aspera of femur and lateral intermuscular septum	Lateral head of fibula and lateral tibial condyle	Extension of thigh (long head), flexion of leg, lateral rotation of flexed leg	Sciatic nerve: tibial branch to long head (L5-S3), common fibular branch to short head (L5-S2)
Psoas	Bodies and transverse processes of lumbar vertebrae	Lesser trochanter of femur	Flexion of thigh and slight adduction of the thigh of free limb	L1-L4

TABLE 6.3

Muscles for Hip Abduction Analysis

Muscle	Origin	Insertion	Action	Innervation
Gluteus medius	Lateral surface of ilium between anterior and posterior gluteal line	Greater trochanter	Abduction, medial rotation and flexion (anterior fibers), and lateral rotation and extension (posterior fibers) of thigh	Superior gluteal nerves L5-S1
Pectineus	Superior ramus of pubis	Femur just distal to lesser trochanter	Flexion and adduction of thigh	Femoral nerve, possibly obturator or accessory obturator nerve or both
Adductor longus	Pubic tubercle	Medial lip of linea aspera of femur	Adduction and flexion of thigh	Obturator nerve
Gracilis	Inferior ramus of pubis, ramus of ischium	Medial surface, proximal end of tibia just distal to medial condyle	Adduction of thigh, flexion of leg, medial rotation of flexed leg	Obturator nerve
Adductor brevis	Body and inferior ramus of pubis	Pectinial line, proximal part of linea aspera of femur	Adduction and flexion of thigh	Obturator nerve
Adductor magnus	Inferior ramus of pubis, ramus of ischium, ischial tuberosity	Linea aspera (anterior fibers), adductor tubercle of femur (posterior fibers)	Adduction, flexion (anterior fibers), and extension of thigh (posterior fibers)	Obturator nerve (anterior fibers), sciatic nerve (posterior fibers)
Tensor fasciae latae	Iliac crest posterior to anterior superior iliac spine	Iliotibial tract	Flexion, medial rotation, and abduction of thigh	Superior gluteal nerve L4-S1
Quadratus lumborum	Medial part of iliac crest	12th rib, lower lumbar vertebrae	Lateral flexion of vertebral column; fixes last rib to form stable base for contraction of diaphragm	Branches from T12 and L1-L3
Piriformis	Sacrum (pelvic surface)	Greater trochanter of femur	Lateral rotation of thigh of extended hip, abduction of thigh when the thigh is flexed	Ventral rami S1 and S2

TABLE 6.4

Muscles for Trunk Flexion Analysis

Muscle	Origin	Insertion	Action	Innervation
Rectus abdominis	Costal cartilage 5-7 and xiphoid process	Pubic symphysis and pubic crest	Flexes trunk	Ventral rami T6-T12
Internal oblique	Inferior border of ribs 10-12, linea alba, and pubis	Thoracolumbar fascia, iliac crest, lateral inguinal ligament	Flexes and rotates trunk	Ventral rami T6-L1
External oblique	External aspects of ribs 5-12	Anterior iliac crest, linea alba, pubic tubercle	Flexes and rotates trunk	Ventral rami T6-T12 and subcostal nerve
Transverse abdominis	Internal aspect of costal cartilage 7-12, thoracolumbar fascia, iliac crest, lateral inguinal ligament	Linea alba, pubis, pubic crest	Supports abdominal viscera and increases intra-abdominal pressure	Ventral rami T6-L1
Erector spinae Iliocostalis lumborum	Iliac crest, sacrum	Lower borders of lowest seven ribs	Bilateral action: extension of trunk	Dorsal rami of spinal nerves in corresponding area
Longissimus thoracis	Intermediate part of extensor aponeurosis	Lower 10 ribs and adjacent vertebral transverse processes	Unilateral action: lateral trunk flexion ipsilaterally	
Psoas	Bodies and transverse processes of lumbar vertebrae	Lesser trochanter of femur	Flexion of thigh and slight adduction of the thigh of free limb	L1-L4

Balance Testing

Balance is considered the ability of a person to maintain equilibrium in an upright posture by maintaining the center of gravity within a base of support. Balance can also refer to the ability to control body movements. Some have used the terms "balance," "proprioception," and "kinesthesia" interchangeably, but each of these terms has a specific meaning. Although not clearly defined, proprioception and kinesthesia generally refer to the ability of the central nervous system to sense position and movements of a given body segment in space (Sherrington 1906; Madey et al. 1993;

Fredericks 1996; Gillquist 1996). Proprioception and kinesthesia are harder to test than balance, and such testing often requires sophisticated balance-type equipment.

Because there are multiple factors that determine balance, the assessments are done in weight bearing. This allows incorporation of the entire kinetic chain (trunk, hips, knees, and ankles). Since proprioception and kinesthesia often require expensive equipment, this chapter looks more closely at balance as it relates to the lower extremity.

SINGLE-LEG STANCE TEST

▶ **Purpose:** To assess the ability to balance on a single leg.

▶ **Equipment:** Flat, nonslip surface; stopwatch; paper and pencil.

Procedure

1. Ask the client to remove shoes and place hands on hips or arms across the body.
2. Instruct the client to lift the nonsupporting leg off the ground. Ensure that the nonsupporting leg is not held against the supporting leg.
3. Give the client a 1 min practice session.
4. Start the stopwatch when the nonsupported foot leaves the ground.
5. For safety, stand behind the client in case of loss of balance.
6. End the test when any of the following occurs: Hands come off the hips, supporting foot swivels or moves (hops) in any direction, or nonsupporting foot touches supporting leg.
7. Use the time the client remains balanced as the score.

Analysis and Interpretation of Data

The best of three attempts is recorded. Side-to-side assessments are made for comparison.

Statistics

Interrater reliability for repeat testing of 30 patients referred to physical therapy had kappa levels of 0.905.

STATIC BALANCE: STORK TEST

▶ **Purpose:** To assess the ability to balance on the ball of the foot.

▶ **Equipment:** Flat, nonslip surface; stopwatch; paper and pencil.

Procedure (Anderson et al. 2000)

1. Instruct the client to remove shoes and place hands on hips.
2. Have the client place the nonsupporting foot against the inside knee of the supporting leg.
3. Give the client 1 min to practice the test.
4. Ask the client to raise the heel and to balance on the ball of the foot.
5. Start the stopwatch as the heel is raised from the floor.
6. For safety, stand behind the client in case he or she loses balance or falls backward.
7. End the test when any of the follow occurs: Hands come off the hips, supporting foot swivels or moves (hops) in any direction, nonsupporting foot loses contact with the knee, or the heel of the supporting foot touches the floor.
8. Use the time the client remains balanced as the score.

Analysis and Interpretation of Data

The best of three attempts is recorded. Side-to-side assessments are made for comparison.

Statistics

Several studies have assessed reliability of the stork test. Johnson and Nelson (1986) reported reliability values of 0.87 when testing sessions were performed on different days. Atwater and colleagues (1990) reported high to moderate reliability coefficients when right and left feet were used both with eyes open and with eyes closed. Ageberg and colleagues (1998) assessed variables that had correlations to the single-leg stance test and obtained high correlations with performance of the single-leg hop test ($r = 0.73$ to 0.95).

FOUR SQUARE STEP TEST

▶ **Purpose:** This test, described by Whitney and colleagues (2007), is designed to identify balance and vestibular dysfunction. The test incorporates stepping backward, forward, and sideways quickly; these movements can be difficult in people with vestibular disorders.

▶ **Equipment:** Flat, nonslip surface; four canes; stopwatch.

Procedure

1. Use standard canes (Dite & Temple 2002) or T-shaped canes.

2. Lay the four canes on the ground at 90° angles to each other (like a plus sign).

3. Ask the client, with shoes on, to stand in one square, directly facing the square in front.

4. Instruct the client to travel clockwise around the "plus sign" by moving forward, then to the right, then backward, then to the left.

5. Instruct the client to reverse the path and repeat the sequence in a counterclockwise direction.

6. Ensure that both feet make contact with the floor in each square.

7. Instruct the client to try to complete the sequence as fast as possible without touching the canes.

8. Tell the client to face forward during the entire sequence if possible.

9. Record the time to complete the sequence.

Analysis and Interpretation of Data

The best of three attempts is recorded.

Statistics

The four square step test has been found to have inter-rater reliability at $r = 0.99$. With a cutoff score of more than 15 s, the test has a sensitivity score of 89%; for non-multiple fallers, it has a specificity of 85% with a positive predictive value of 86% for detecting a history of falls among community-living adults (Dite & Temple 2002).

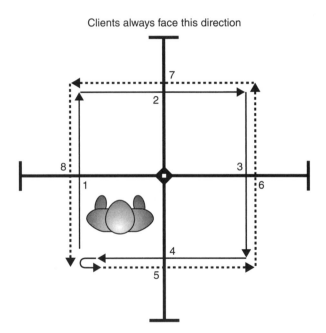

Clients always face this direction

SIDE-STEP TEST

▶ **Purpose:** To assess balance by determining the distance the client can sidestep. Sidestepping requires considerable balance and coordination.

▶ **Equipment:** Athletic tape for a starting line and a line that runs 10 m perpendicular to the starting line.

Procedure

1. Instruct the client to stand in the starting position with legs and feet together on the starting line; in principle, the feet are in contact in this position.
2. Ask the client to sidestep five steps, attempting to make the steps as wide as possible.
3. Instruct the client not to support the body with the arms and not to jump.

Analysis and Interpretation of Data

- Measure the maximum side-step length and divide the total distance moved by the number of side steps taken.
- To standardize the maximum side-step length, you can divide by the leg length (the distance between the anterior superior iliac spine and the medial malleolus).

Statistics

Fujisawa and Takeda (2006) studied correlations between the side-step test and several commonly used measures of balance and mobility in 28 hemiplegic subjects (17 with left-sided impairment and 11 with right-sided impairment). They found a high linear correlation between maximum side-step length and maximum walking speed, and between maximum side-step length and stride length ($P < 0.01$). Pearson's product–moment correlation coefficients between 0.84 and 0.89 were obtained for this test. A nonlinear, positive correlation was observed between maximum side-step length and one-footed standing duration. Although actual reliability measures were not presented, Fujisawa and Takeda (2006) reported high test–retest reliability, which they suggested should help increase the popularity of the test in clinics.

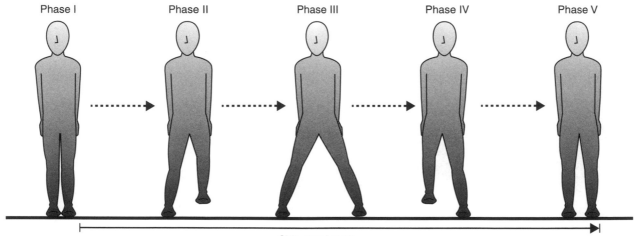

Side-step length

STAR EXCURSION BALANCE TEST

▶ **Purpose:** To determine balance and postural control in a dynamic test. This functional test is purported to detect functional performance deficits associated with lower extremity pathology in either healthy or impaired clients. It uses a series of lower extremity reaching tasks in up to eight directions that challenge the client's postural control, strength, range of motion, and proprioceptive abilities.

▶ **Equipment:** Athletic tape placed in diagonal patterns, or functional testing grid.

Procedure

1. Instruct the client to reach with one lower extremity while balancing on the opposite.
2. Ask the client to reach in eight different directions at 45° increments from the center of the grid.
3. Instruct the client to reach to the farthest possible point with the distal-most part of the foot.
4. Tell the client to reach as far as possible without using the reaching leg to provide support in the maintenance of the upright posture.

5. Instruct the client to return to the upright position after each reach while maintaining equilibrium.
6. Instruct the client not to move the stance foot from its original position.
7. Perform the test for both the dominant and the nondominant limb.

Analysis and Interpretation of Data

■ This test is scored by the distance reached. A larger score indicates better balance, range of motion, strength, and proprioception abilities.

■ If a difference in scores exists, the ability to reach farther with a limb requires a combination of better balance, strength, and motion than the contralateral limb.

Statistics

■ Reliability studies for the Star Excursion Balance Test (SEBT) or Lower Extremity Functional Reach Test are outlined in table 7.1 on page 117.

Anterior reach.

Medial reach.

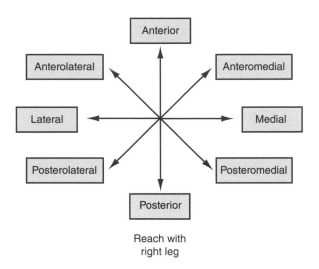

Reach with
right leg

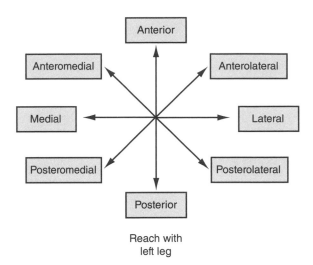

Reach with
left leg

- Hertel and colleagues (2000) originally recommended nine trial sessions for this test since they identified learning effects in four of the eight directions, with the longest excursion distances occurring during trials 7 through 9. More recently Robinson and Gribble (2008) have reported that fewer trials are needed. Their study of 20 participants showed that for the majority of reach directions, maximal excursions occurred within the first four practice trials. Therefore their recommendation is to simplify the test by decreasing practice trials from six to four.

- Olmsted and colleagues (2002) found that subjects with unilateral chronic ankle instability reached significantly shorter distances on their involved limb compared to their uninvolved limb and compared to the side-matched limbs of a control group. In their analysis, reach distances from all eight directions of the SEBT were averaged together.

- Gribble and Hertel (2003) reported that reach deficits were exacerbated in subjects with and without chronic ankle instability after lower extremity fatiguing exercise.

- Recently Hertel and colleagues (2006) simplified the SEBT while assessing changes in subjects with chronic ankle instability. They determined that subjects with chronic ankle instability reached significantly shorter distances in the anteromedial, medial, and posteromedial directions when balancing on their involved limb compared to their uninvolved limb and compared to the side-matched limbs of controls. The authors suggest that the anteromedial, medial, and posteromedial reach directions may be used clinically to test for functional balance and reach deficits in lieu of testing all eight directions as in the test originally described.

- Plisky and colleagues (2006) recently found, using a logistic regression model, that high school basketball players with an anterior right–left reach distance difference greater than 4 cm were 2.5 times more likely to sustain a lower extremity injury. Girls with a composite reach distance less than 94% of their limb length were 6.5 times more likely to sustain a lower extremity injury.

FUNCTIONAL REACH TEST

▶ **Purpose:** To determine a client's functional ability to reach with the upper extremity while maintaining trunk stability.

▶ **Equipment:** Yardstick or tape measure.

Procedure

1. Instruct the client to reach as far as possible (shoulders at 90° of flexion, elbows fully extended).

2. Use a yardstick mounted on the wall at the level of the acromion to measure reach distance.

3. Measure reach distance as the displacement of the finger between initial position and end position.

Analysis and Interpretation of Data

The test is scored by the distance reached.

Statistics

Johnsson and colleagues (2002) reported that the correlation between functional reach test scores in elderly persons and the displacement of center of pressure is low, indicating that this test may not reflect the stability limit. Only 15% ($r^2 = 0.15$) of the variation in center of pressure displacement could be explained by how far a subject could reach, leaving 85% of the variation to other factors. This means that the task of reaching may be influenced by other factors such as movement of the trunk.

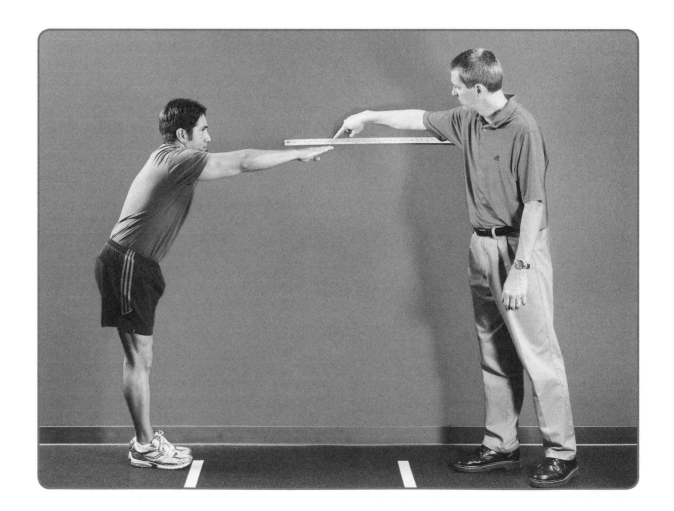

ROMBERG TEST

▶ **Purpose:** To assess balance.

▶ **Equipment:** No equipment is required.

Procedure (Anderson et al. 2000; Starkey & Ryan 2003)

1. Instruct the client to stand with feet together, arms at the side, and eyes closed while maintaining balance.

2. You can use variations of this test in which the client holds the arms at 90° of abduction, tilts the head backward, stands on the toes, stands on a single leg, or performs the finger-to-nose test (touching fingers to nose).

Analysis and Interpretation of Data

Any postural sway would indicate loss of balance. Gross unsteadiness may indicate cerebellar dysfunction.

Statistics

There are no known reliability or validity statistics for this assessment.

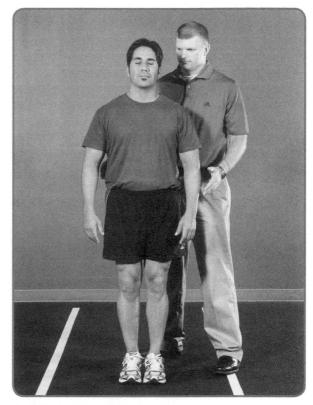

Variation of Romberg test.

Variation of Romberg test.

Variation of Romberg test.

TANDEM WALKING

▶ **Purpose:** To assess balance.

▶ **Equipment:** Gym floor with straight line on the edge or athletic tape placed in a straight line of at least 10 yd (9 m).

Procedure (Starkey & Ryan 2003)

1. Instruct the client to stand with the feet straddling a straight line.
2. Ask the client to walk heel to toe along the straight line for approximately 10 yd (9 m).
3. Instruct the client to return to the starting position by walking backward.

Analysis and Interpretation of Data

Any postural sway would indicate loss of balance. Gross unsteadiness may indicate cerebellar or inner ear dysfunction.

Statistics

There are no known reliability or validity statistics for this assessment.

TINETTI TEST

▶ **Purpose:** To determine a client's gait and balance.

▶ **Equipment:** An open hallway or room and a standard chair, also any assistive device that the client may require.

Procedure (Tinetti 1986)

1. For the balance portion of the examination, have the client seated in a hard armless chair and ask him or her to rise and stand.

2. Score balance according to the Tinetti Assessment Tool for Balance on page 277 in the appendix. (For your convenience, the assessment tool is also included on the DVD that accompanies this book.)

3. For the gait portion of the examination, instruct the client to stand and walk down a hallway or across a room at a "usual pace" and then return at a "rapid but safe" pace.

4. Score the gait according to the Tinetti Assessment Tool for Gait on page 278 in the appendix. (For your convenience, the assessment tool is also included on the DVD that accompanies this book.)

Analysis and Interpretation of Data

Scoring uses a 3-point ordinal scale range of 0 to 2. A score of 0 equates to "most impaired"; a score of 2 equates to client independence. Scores are added to form three different measures: (1) overall gait assessment, (2) overall balance assessment, and (3) gait and balance score. Maximum score for gait is 12 points, while maximum score for the balance test is 16. A maximum score for gait and balance equals 28 points. A client who scores below 19 is at a high risk for falls. Clients who score in the range of 19 to 24 have a risk for falls.

Statistics

None available.

BALANCE ERROR SCORING SYSTEM

▶ **Purpose:** To assess postural control with clients who may have postural difficulties. The system was originally developed to assess athletes with a potential mild head injury in sideline situations during sport.

▶ **Equipment:** Firm surface and soft surface (foam), stopwatch.

Procedure

The test is composed of six 20 s conditions including a double-leg, single-leg, and tandem (heel-to-toe) stance performed on a firm surface and then repeated on a foam type of surface.

1. Instruct the client to perform a double-leg stance on the firm surface, then a single-leg stance, and then a tandem stance.
2. Conduct the same testing in the same order on the foam surface (46 $\times$ 43 $\times$ 13 cm^2 block of medium-density foam pad from Airex [Alcon Airex, Aargau, Switzerland]).
3. Instruct the client to stand with hands on the iliac crest, head up, facing forward with eyes closed.
4. Count the following as errors:
   - Client opens eyes.
   - Client steps, stumbles, or falls out of the test position.
   - Client removes hands from hips.
   - Client moves hips into more than 30° of flexion or abduction.
   - Client lifts toes or heels from test surface.
   - Client remains out of test position for longer than 5 s.
5. Instruct the client to make any necessary adjustments if he or she loses balance and to return to position as quickly as possible.
6. Instruct the client to maintain balance and not commit any of the balance errors listed.

Analysis and Interpretation of Data

■ A stopwatch is used to time 20 s sessions while errors are counted.

■ A minimum score of zero errors and a maximum score of 10 errors are possible for each of the six testing conditions. To calculate a total score, scores from the subcategories are added.

■ Higher scores are thought to be associated with decreased postural control.

Statistics

■ Several studies have presented intratester reliability coefficients for this test, ranging from 0.78 to 0.96 (Riemann & Guskiewicz 2000; Valovich et al. 2004).

■ Riemann and partners found significant correlations between the Balance Error Scoring System (BESS) and sophisticated force platform measurements of postural sway in normal subjects (Riemann & Guskiewicz 2000). This simply means that the BESS can be used in clinical situations with hopes of obtaining results similar to those that can be acquired with the use of expensive computerized equipment. Expanding on this concept, Riemann and Guskiewicz (Riemann & Guskiewicz 2000) assessed the efficacy of the BESS for detecting acute postural stability changes following mild head injury. Subjects performed the BESS on days 1, 3, 5, and 10 following their injury. The BESS demonstrated significantly higher postural instability in subjects with mild head injury compared to controls through day 3 postinjury.

■ Onate and colleagues (2007) assessed the extent to which the environment affected BESS scores. This test was developed to assess postural stability on the field, and the authors wanted to see if scores would be different across other settings such as a laboratory environment. They found significant group mean differences between testing environments for the single-leg foam stance ($P = 0.001$). Scores in the uncontrolled sideline environment (7.33 ± 2.11 errors) were higher than in the controlled clinical environment (5.19 ± 2.16 errors). Additionally, medium to large effect sizes (0.53 to 1.03) were found for single-leg foam, tandem foam, and total BESS scores, with relative increases (worse scores) of 30% to 44% in the sideline environment compared with the clinical environment.

MULTIPLE SINGLE-LEG HOP STABILIZATION TEST

▶ **Purpose:** To assess postural control and balance during functional performance with repetitive hopping.

▶ **Equipment:** Open, nonslip testing surface; athletic or colored tape.

Procedure (Riemann et al. 1999)

1. Set up the floor pattern consisting of 11 square boxes of tape; each box is 2.5 cm square, and intertape distances are as shown in the diagram.
2. Describe the testing sequence to the client.
3. Allow the client to perform several practice hop stabilization trials on each leg prior to testing.
4. Have the client stand with the test lower extremity completely covering the start box, the nontest lower extremity flexed slightly at the hip and knee to avoid ground contact, and hands on bilateral iliac crests.
5. Allow the client to look briefly at the target location before hopping to tape mark 1.
6. Instruct the client that upon landing, he or she must control balance to remain in single-leg stance with hands on iliac crests and head remaining level and facing forward.
7. Once the client establishes control, begin counting 5 s aloud to mark the beginning of the balance period. During this period the client must maintain a stable position, looking forward, without touch-

ing down or moving the nontest leg into excessive (>30°) flexion, abduction, or extension.

8. At the end of 5 s, allow the client again to look at the target location and hop to the next tape mark, landing in the standardized position.
9. Determine the success of each landing and balancing period using criteria listed in the next section.
10. Have the client perform the hop course on the noninvolved leg; then, after a rest period (client is fully recovered), conduct the testing with the involved leg. The score is interpreted as described next.

Analysis and Interpretation of Data

■ Committing one of the errors (landing or balance) during a period counts as a failure for that entire period.

Landing errors	Balance errors
Not covering tape mark	Touching down with nontest leg
Stumbling on landing	Nontest leg touching test leg
Foot not facing forward with 10° of inversion or eversion	Nontest leg moving into excessive flexion, extension, or abduction
Hands off hips	Hands off hips

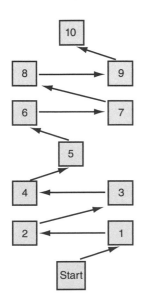

Intertape distances

Body height (cm)	Diagonal distance (cm)	Adjacent distance (cm)
150-159.9	70	49
160-169.9	74	53
170-179.9	79	56
180-189.9	83	59
190-199.9	88	62
200-209.9	92	66

(continued)

(continued from previous page)

- At the conclusion of the test, 10 error points are given for each period in which a landing error is committed, and 3 error points are given for each period in which a balance error is committed.

- The sum of the error points is the total score.

- Scores for 30 recreationally active subjects (19 males and 11 females) with mean age of 21.23 ± 2.9 years (Riemann et al. 1999) were as follows:

  - Balance score: 7.3 ± 5.9 (mean ± SD).

  - Landing score: 43.7 ± 23.3 (mean ± SD).

- Caution: Several studies have demonstrated, in terms of the error of >30° of flexion at the knee, specifically with females, that individuals landing with less knee flexion have an increased risk of knee injury.

Statistics

- Landing score: intraclass correlation coefficient (ICC) = 0.92; SEM 0.57 (Riemann et al. 1999).

- Balance score: ICC = 0.70 to 0.74; SEM 0.55 (Riemann et al. 1999).

- A modified version of this test was shown to be reliable in demonstrating functional performance deficits in patients with chronic ankle instability (Eechaute et al. 2008).

  - Unstable ankles (n = 17): ICC = 0.91 to 0.97 (Eechaute et al. 2008).

  - Healthy ankles (n = 29): ICC = 0.87 (Eechaute et al. 2008).

TABLE 7.1

Reliability of the Star Excursion Balance Test
or Lower Extremity Functional Reach Test

Authors	Intratester reliability	Intertester reliability
Austin & Scibek 2002	0.96	0.93
Hertel et al. 2000	0.85 to 0.96	0.81 to 0.93
Kinzey & Armstrong 1998	0.67 to 0.87	Not tested
Loudon et al. 2002	0.83	Not tested
Manske & Andersen 2004	0.94 to 0.98	Not tested

Aerobic Testing

Aerobic testing utilizes various test procedures that determine a client's fitness levels. Fitness encompasses many variables, including cardio-respiratory levels. The cardiovascular response to these tests is generally to deliver oxygen and other nutrients to the muscles being used. One of the most common methods of determining one's fitness level is to measure what is known as $\dot{V}O_2$max. $\dot{V}O_2$max is defined by McArdle and colleagues (2001) as a person's capacity for aerobic resynthesis. This equates to a person's ability to sustain a level of high-intensity exercise, which is typically for as long as 4 to 6 min in most healthy individuals.

The tests described in this chapter require clients to perform exercise to a maximal level for an extended period of time. The client will be very fatigued during and especially after these tests. Safety is of the highest concern during and following the tests. These tests should in most cases be discontinued if a client reports any symptoms such as pain, light-headed-ness, dizziness, nausea, or inability to continue the testing session due to extreme fatigue.

ONE-MILE WALK TEST

▶ **Purpose:** To determine the client's level of cardiorespiratory fitness ($\dot{V}O_2$max). This test utilizes an alternate method (other than using respirators or expiratory devices) to determine $\dot{V}O_2$max, which is the rate at which oxygen is consumed during a given activity.

▶ **Equipment:** Measured track or course or treadmill, stopwatch; heart rate (HR) monitor is optional.

Procedure (American Heart Association 1990)

1. Instruct the client to walk along the measured testing surface as fast as he or she can without running.

2. If HR monitor is not available, measure the client's pulse rate manually and record HR in beats per minute immediately upon completion of the test.

3. Record the elapsed time to complete the walk to the nearest second.

Analysis and Interpretation of Data

Calculation of the estimated $\dot{V}O_2$max is as follows:

$$(ml \cdot kg^{-1} \cdot min^{-1}) = 132.85 - (0.077 \times \text{Body weight in pounds}) - (0.39 \times \text{Age in years}) + (6.32 \times \text{Gender } [0 = F; 1 = M]) - (3.26 \times \text{Elapsed time in minutes}) - (0.16 \times \text{HR in beats per minute}).$$

Table 8.1 on page 126 and table 8.2 on page 127 give percentile scores (times) based on age for girls and boys, respectively.

Statistics

There are no known reliability or validity data for this assessment.

ROCKPORT WALK TEST

▶ **Purpose:** To determine the client's level of cardiorespiratory fitness ($\dot{V}O_2$max). This test utilizes an alternate method (other than using respirators or expiratory devices) to determine $\dot{V}O_2$max, which is the rate at which oxygen is consumed during a given activity.

▶ **Equipment:** Stopwatch; 1-mile (1.6 km) track, smooth and level; HR monitor (optional).

Procedure (American College of Sports Medicine 1995; D'Alonzo et al. 2006)

1. Instruct the client to walk as fast as possible over the 1-mile course.

2. Immediately after the client has completed the 1-mile distance, take his or her pulse either with a pulse monitor (preferably) or manually.

3. Measure the time taken to complete the 1-mile course.

Analysis and Interpretation of Data

A $\dot{V}O_2$max score for females can be calculated using the following equation:

$\dot{V}O_2$ = 139.168 – (0.388 × Age) – (0.077 × Weight in pounds) – (3.265 × Walk time in minutes) – (0.156 × HR).

For males, add 6.318 to the preceding equation:

$\dot{V}O_2$ = 132.853 – 0.0769 x weight (lbs) – 0.3877 x age + 6.318 x gender value (0 for females, 1 for males) – 3.249 x 1 –mile walk time – 0.1565 x heart rate

Normative values for this test are listed in table 8.3.

Statistics

■ Concurrent validity between the Rockport Walk Test and treadmill testing: r = .88 (Kline et al. 1987).

■ Test–retest reliability: r = .98 (Kline et al. 1987).

Note

This test is more suitable for both males and females of poor fitness who would not be able to complete a similar distance run test; it is too easy for highly fit individuals.

CHESTER STEP TEST

▶ **Purpose:** The Chester Step Test is a submaximal test of cardiovascular function that utilizes multiple stages while the HR and exertion levels are continuously monitored.

▶ **Equipment:** A metronome for audio step frequency and a 30 cm step, graphical datasheet on page 279 in the appendix. (For your convenience, the graphical datasheet is also included on the DVD that accompanies this book.)

Procedure (Sykes 1995)

1. Review testing procedures with the client before beginning the test.
2. Instruct the client to begin with a step rate of 15 steps per minute for 2 min (level 1).
3. Record both heart rate and perceived exertion.
4. Then increase the step rate to 20 steps per minute for another 2 min (level 2).
5. Again record the heart rate and perceived exertion.
6. Continue testing in this progressive manner until the client reaches an exercising HR of 80% of his or her predicted maximum (220 – age).

Analysis and Interpretation of Data

■ The level that the client last finished is used for scoring. Therefore if the client was in the middle of level 4 when he or she had to stop, level 3 would be used for the score.

■ A maximum test duration of 10 min is given (i.e., level 5).

■ To predict aerobic capacity, plot the exercise HRs on the prepared graphical datasheet, drawing a visual line of best fit between the data points; project the line up to maximum HR and estimate the corresponding aerobic capacity (ml $O_2 \cdot kg^{-1} \cdot min^{-1}$) from the x-axis.

Statistics

Sykes and Roberts (2004) found high correlations (r = 0.92) between the standard treadmill $\dot{V}O_2$max aerobic capacity test and the Chester Step Test.

Note

The client should be stopped if he or she displays signs of overexertion or distress or if the rating of perceived exertion is greater than 15.

20-METER SHUTTLE RUN TEST

▶ **Purpose:** To determine the client's level of cardio-respiratory fitness.

▶ **Equipment:** Calibrated cassette tape drive or CD version of the 20 m shuttle run test (CD version is preferable due to tape stretch with cassette tape; available for purchase from Australian Sports Commission), gym floor with tape or markers at 20 m.

Procedure

1. Set up the course, using tape or markers to indicate the start line and the 20 m line.
2. Inform the client of the testing procedures. When the prerecorded tape begins, the client will run back and forth on the course and touch the 20 m line at the same time the tape emits a sound signal (the frequency of the signal increases in such a way that running speed is increased by 0.5 km/h each minute from a starting speed of 8.5 km/h).
3. Start the tape and instruct the client to begin the test.
4. Stop the test when the client is no longer able to follow the set pace.

Analysis and Interpretation of Data

The last stage number announced on the tape or the equivalent maximal aerobic speed is used as the $\dot{V}O_2$max index.

Statistics

Leger and colleagues (1988) found the 20 m shuttle run test to be reliable in children ($r = 0.89$) and adults ($r = 0.95$).

MULTISTAGE FITNESS TEST (20-METER SHUTTLE RUN, YO-YO TEST)

▶ **Purpose:** To determine the client's level of cardiorespiratory fitness.

▶ **Equipment:** Calibrated cassette tape drive or CD version of the 20 m shuttle run test (CD version is preferable due to tape stretch with cassette tape; available for purchase from Australian Sports Commission), flat nonslip testing surface, two cones to mark distances, stopwatch.

Procedure (Leger & Gadoury 1989; Ramsbottom et al. 1988; Shvartz & Reibold 1990)

1. Mark a 20 m distance with one cone at each end.
2. Instruct the client to carefully listen to the tape or CD and advise the client of the testing criterion of always placing a foot on or behind the 20 m mark.
   - The tape or CD will emit a beep at the time the client is required to be at the 20 m mark.
   - The client must try to be at the opposite end of the 20 m track by the time the next beep sounds.
   - The client's running speed will have to gradually increase because the time interval between beeps decreases after approximately each minute.
3. Start the tape or CD and instruct the client to begin the test.
4. Warn the client if he or she is unable to reach the 20 m mark in time for the beep; terminate the test when the client is unable to reach the 20 m mark twice in succession.

Analysis and Interpretation of Data

- The client's score is the level and number of shuttles immediately previous to the beep on which he or she was eliminated.
- More detailed instructions for calculations of $\dot{V}O_2max$ according to the client's score on the test are given on the tape or CD, but scores can generally be calculated from the following formula:

   $$\dot{V}O_2max = 3.46 \cdot [1 \cdot Level + (Shuttles / [Level \cdot .4325 + 7.0048])] + 12.2.$$

- Normative values can be estimated from table 8.4 on page 128.
- Rough estimates of percentages of category of fitness are given in table 8.5 on page 129. $\dot{V}O_2max$ for athletes in various sports are as follows:
   - Average: Females ranged from 35 to 43 and males ranged from 44 to 51 ml · kg^{-1} · min^{-1}.
   - Above average: Females ranged from 44 to 48 and males ranged from 52 to 56 ml · kg^{-1} · min^{-1}.

Statistics

This test has been found to be a sufficiently accurate estimate of aerobic power (Brewer et al. 1988; Leger & Lambert 1982).

1.5-MILE RUN TEST

▶ **Purpose:** To determine the client's level of cardio-respiratory fitness.

▶ **Equipment:** Measured track or course (treadmill is a possible alternative), stopwatch.

Procedure

1. Instruct the client to run the 1.5-mile (2.41 km) course as fast as he or she is capable of running.
2. Record the time in minutes and to the nearest tenth of a second from the initiation of the test until the 1.5-mile distance is completed.

Analysis and Interpretation of Data

■ To estimate the $\dot{V}O_2max$, use the following formula ($\dot{V}O_2$ is in $ml \cdot kg^{-1} \cdot min^{-1}$ and time is in minutes):
$$\dot{V}O_2max = 3.5 + 483 / (\text{Time to run 1.5 mi or 2.41 km}).$$

■ NCAA Division I male basketball players: $9.43 \pm 1:06$ (Latin et al. 1994).

■ Normative data can be found in table 8.6 and table 8.7 on page 129.

Statistics

There are no known reliability or validity data for this assessment.

12-MINUTE RUN TEST

▶ **Purpose:** To determine the client's level of cardio-respiratory fitness. This is an alternative assessment that allows for use of a set period of time and scoring according to distance versus marking a set distance and using time as the assessment.

▶ **Equipment:** Measured track or course (treadmill is possible alternative), stopwatch.

Procedure (Hoffman 2006)

1. Review testing procedures with the client before beginning the test. Inform the client that he or she is allowed to walk or rest during the test if necessary, but that this will affect the score on the assessment.
2. Instruct the client to run continuously for a 12 min period over a set course marked for standard distance (e.g., track). Advise the client to cover as much distance as possible over this 12 min period.

3. Start the stopwatch as soon as the client begins the test and stop it at the 12 min mark.

Analysis and Interpretation of Data

■ The distance covered from the start of the assessment to where the client stops at the 12 min mark is measured for the client's score.

■ To estimate the $\dot{V}O_2max$, use the following formula:
$$\dot{V}O_2max = 0.0268 (\text{Distance covered}) - 11.3;$$
$\dot{V}O_2$ is in $ml \cdot kg^{-1} \cdot min^{-1}$ and distance covered is in meters.

Statistics

None available.

TABLE 8.1

Percentile Scores Based on Age and Test Scores for One-Mile Run/Walk for Girls (min:s)

PERCENTILE	AGE											
	6	7	8	9	10	11	12	13	14	15	16	17+
100	8:36	8:04	8:00	6:11	6:26	7:07	6:22	5:42	5:00	5:51	5:58	6:20
95	10:06	9:30	9:10	8:21	8:07	8:06	7:35	7:21	7:20	7:25	7:26	7:22
90	10:29	10:05	9:45	9:07	8:49	8:40	8:00	7:49	7:43	7:52	7:55	7:58
85	11:20	10:36	10:02	9:30	9:19	9:02	8:23	8:13	7:59	8:08	8:23	8:15
80	11:37	10:55	10:20	10:03	9:38	9:22	8:52	8:29	8:20	8:24	8:39	8:34
75	12:00	11:17	10:55	10:22	10:08	9:44	9:15	8:49	8:36	8:40	8:50	8:52
70	12:12	11:25	11:20	10:45	10:19	10:04	9:36	9:09	8:50	8:55	9:11	9:15
65	12:20	11:45	11:38	10:58	10:42	10:24	10:05	9:30	9:09	9:09	9:25	9:33
60	12:31	12:20	11:53	11:13	10:52	10:42	10:26	9:50	9:27	9:23	9:48	9:51
55	12:45	12:39	12:10	11:32	11:00	11:00	10:44	10:07	9:51	9:37	10:09	10:08
50	13:12	12:56	12:30	11:52	11:22	11:17	11:05	10:23	10:06	9:58	10:31	10:22
45	13:56	13:21	12:46	12:13	11:40	11:36	11:23	10:57	10:25	10:18	10:58	10:48
40	14:14	13:44	13:07	12:24	11:58	12:00	11:47	11:20	10:51	10:40	11:15	11:05
35	14:45	14:04	13:31	12:48	12:08	12:21	12:01	11:40	11:10	11:00	11:44	11:20
30	15:09	14:32	13:56	13:19	12:30	12:42	12:24	12:00	11:36	11:20	12:08	12:00
25	15:27	14:55	14:21	13:44	13:00	13:09	12:46	12:29	11:52	11:48	12:42	12:11
20	16:10	15:12	14:53	14:07	13:29	13:44	13:35	13:01	12:18	12:19	13:23	12:40
15	16:45	16:00	15:19	14:57	14:00	14:16	14:12	14:10	12:56	13:33	14:16	13:03
10	17:36	16:35	15:45	15:40	14:30	14:44	14:39	14:49	14:10	14:13	16:03	14:01
5	19:00	17:27	16:55	16:58	15:43	16:07	16:00	16:10	15:44	15:17	18:00	15:14
0	21:40	22:19	20:40	24:00	24:00	21:02	24:54	20:45	20:04	24:07	21:00	28:50

The President's Challenge Physical Activity and Fitness Awards Program, a program of the President's Council on Physical Fitness and Sports, U.S. Department of Health and Human Services.

TABLE 8.2

Percentile Scores Based on Age and Test Scores for One-Mile Run/Walk for Boys (min:s)

	AGE											
PERCENTILE	**6**	**7**	**8**	**9**	**10**	**11**	**12**	**13**	**14**	**15**	**16**	**17+**
100	6:18	7:41	6:30	6:50	6:24	6:29	6:03	5:40	4:30	4:42	4:49	4:46
95	8:54	8:31	8:00	7:48	7:10	6:56	6:43	6:25	6:01	5:50	5:40	5:35
90	9:41	5:56	8:28	8:14	7:39	7:17	6:57	6:39	6:13	6:07	5:56	5:57
85	10:15	9:22	8:48	8:31	7:57	7:32	7:11	6:50	6:26	6:20	6:08	6:06
80	10:32	9:43	9:00	8:47	8:08	7:45	7:25	7:00	6:33	6:29	6:18	6:14
75	10:53	10:02	9:23	9:04	8:19	8:00	7:41	7:11	6:45	6:38	6:25	6:23
70	11:17	10:20	9:38	9:12	8:37	8:14	7:56	7:20	6:59	6:48	6:33	6:32
65	11:41	10:34	9:56	9:30	8:59	8:27	8:05	7:29	7:09	6:57	6:44	6:40
60	12:00	10:55	10:15	9:47	9:11	8:45	8:14	7:41	7:19	7:06	6:50	6:50
55	12:20	11:19	10:39	10:07	9:29	9:01	8:25	7:55	7:29	7:16	6:58	6:57
50	12:36	11:40	11:05	10:30	9:48	9:20	8:40	8:06	7:44	7:30	7:10	7:04
45	13:00	11:56	11:27	10:46	10:10	9:46	8:58	8:17	7:59	7:39	7:20	7:14
40	13:39	12:17	11:55	11:03	10:32	10:07	9:11	8:35	8:13	7:52	7:35	7:24
35	14:11	12:50	12:08	11:20	10:58	10:25	9:40	8:54	8:30	8:08	7:53	7:35
30	14:48	13:23	12:30	11:44	11:14	10:54	10:00	9:10	8:48	8:29	8:09	7:52
25	15:12	13:49	12:54	12:08	11:40	11:25	10:22	9:23	9:10	8:49	8:37	8:06
20	15:34	14:16	13:23	12:33	12:15	12:00	10:52	10:02	9:35	9:05	8:56	8:25
15	16:30	15:00	14:10	12:59	13:07	12:29	11:30	10:39	10:18	9:34	9:22	8:56
10	17:25	16:12	14:57	13:52	13:50	13:08	12:11	11:43	11:22	10:10	10:17	9:23
5	18:12	17:43	16:08	15:01	14:47	14:35	13:14	12:47	12:11	11:25	11:49	10:15
0	22:05	21:20	22:40	19:40	23:00	23:32	23:05	24;12	18:10	21:44	20:15	16:49

The President's Challenge Physical Activity and Fitness Awards Program, a program of the President's Council on Physical Fitness and Sports, U.S. Department of Health and Human Services.

TABLE 8.3

Normative Values for the Rockport Walk Test

AGES 30-69 YEARS (MIN:S)		
Rating	**Males ($n = 151$)**	**Females ($n = 150$)**
Excellent	<10:12	<11:40
Good	10:13-11:42	11:41-13:08
High average	11:43-13:13	13:09-14:36
Low average	13:14-14:44	14:37-16:04
Fair	14:45-16:23	16:05-17:31
Poor	>16:24	>17:32
AGES 18-30 YEARS (MIN:S)		
Percentile	**Males ($n = 400$)**	**Females ($n = 426$)**
90%	11:08	11:45
75%	11:42	12:49
50%	12:38	13:15
25%	13:38	14:12
10%	14:37	15:03

Reprinted, by permission, from J.R. Morrow et al., 2005, *Measurement and evaluation in human performance,* 3rd ed. (Champaign, IL: Human Kinetics), 235.

TABLE 8.4

Cardiorespiratory Fitness Classification: $\dot{V}O_2$max (ml $\cdot$ kg^{-1} $\cdot$ min^{-1})

Age (years)	Poor	Fair	Good	Excellent	Superior
WOMEN					
20-29	≤35	36-39	40-43	44-49	50+
30-39	≤33	34-36	37-40	41-45	46+
40-49	≤31	32-34	35-38	39-44	45+
50-59	≤24	25-28	29-30	31-34	35+
60-69	≤25	26-28	29-31	32-35	36+
70-79	≤23	24-26	27-29	30-35	36+
MEN					
20-29	≤41	42-45	46-50	51-55	56+
30-39	≤40	41-43	44-47	48-53	54+
40-49	≤37	38-41	42-45	46-52	53+
50-59	≤34	35-37	38-42	43-49	50+
60-69	≤30	31-34	35-38	39-45	46+
70-79	≤27	28-30	31-35	36-41	42+

Adapted, by permission, from The Cooper Institute, 1997, *The physical fitness specialist manual* (Dallas, TX: The Cooper Institute for Aerobics Research).

TABLE 8.5

Fitness Categories
According to Percentages

Category	Percent (%) of population
Excellent	3
Very good	8
Good	22
Average	34
Fair	22
Poor	8
Very poor	3

TABLE 8.6

Percentile Scores of Police Recruits for 1.5-Mile Run

	PERCENTILE								
Category	90	80	70	60	50	40	30	20	10
1.5-mile run time (min:s)	11:31	12:32	13:14	13:58	14:40	15:20	15:55	16:55	17:00

Data from R.J. Hoffman and T.R. Collingwood, 2005, *Fit for duty,* 2nd ed. (Champaign, IL: Human Kinetics).

TABLE 8.7

Standard Passing Scores (min:s)
for 1.5-Mile Run in Police Department Personnel

	AGE (YEARS)			
Gender	20-29	30-39	40-49	50-59
Females	15:56	15:57	16:58	17:54
Males	12:51	13:36	14:29	15:26

Reprinted, by permission, from J. Hoffman, 2006, *Norms for fitness, performance, and health* (Champaign, IL, Human Kinetics), 76.

Strength and Power Testing

Assessment of strength and power can be multidimensional. Strength, while having potentially multiple definitions, is probably best defined as "the maximal amount of force a muscle or muscle group can generate in a specified movement pattern at a specified velocity of movement" (Knuttgen & Kraemer 1987). Power is the rate of performing work, or force produced divided by the time it takes to produce that force. Therefore, an important component of power production is the amount of time required to produce the force. The only ways to increase power are (1) to increase movement speed and (2) to lift a heavier resistance with the same or greater movement speed than a lighter resistance. One-repetition maximum (1RM) lifts are better indicators of general strength than of power. The reason is that the amount of weight lifted is the maximum amount, and while the weight is generally lifted as fast as possible, the time component is not essential to the force production. The time component of such lifts is typically much slower than the time needed for the production of power.

While debate continues as to optimum training levels for power, it was initially generally accepted, on the basis of comparisons of training loads, that lower loads with higher velocity (specifically 30% of a 1RM load) resulted in the greatest power production (Häkkinen et al. 1985; Toji et al. 1997). Maximal power outputs have been reported to occur at loads ranging from 10% of 1RM (Stone et al. 2003) to 70% 1RM (Baker et al. 2001). While there is considerable variability in the specific load found to maximize power output, it is generally accepted that the client needs an underlying strength base to be able to quickly produce force and therefore power. For this chapter we have chosen to use the traditional definition of strength quoted earlier and to view power as a more explosive, time-dependent quantity. Readers are encouraged to use the listed references as well as other sources to make an informed decision on the best use of these terms in their own settings and situations.

Readers are reminded of a few critical issues that they should address before implementing the types of tests presented in this chapter. Careful planning of the testing sequence, as well as a proper warm-up, is essential. A proper warm-up allows for the safest and most effective utilization of a specific test or testing battery. Proper sequencing of tests helps to ensure safety and to ensure maximum performance when multiple tests are used. Refer to chapters 2 and 3 for issues of warm-up, specific testing sequence, testing purpose, and specific examples of utilization of a testing battery.

CO-CONTRACTION TEST

▶ **Purpose:** To determine the client's ability to cocontract and stabilize bilateral lower extremity agonist and antagonist muscles in a coordinated fashion for time.

▶ **Equipment:** Belt; approximately 122 cm (48 in.) length of rubber tubing with outer diameter of 2.54 cm (Rehab Tubing, Pro Orthopedic Devices, Inc., Tucson, AZ). The tubing is secured to a metal loop on the wall 154 cm (60 in.) above the floor. A semicircle is outlined on the floor with a radius of 244 cm (96 in.) from the metal loop.

Procedure (Lephart et al. 1988)

1. Have the belt strapped around the client's waist and the rubber tubing attached to the belt.
2. Instruct the client to stand facing the wall with toes on the line that forms the semicircle. Tell the client to stretch the tubing 122 cm (48 in.) beyond its recoil length.
3. Instruct the client to complete five wall-to-wall lengths of the 180° semicircle (radius of 96 in. [244 cm] from the metal loop), with the tension applied to the overstretched rubber tubing.
4. Tell the client to use a side-step or shuffle type of motion in each direction and to do the five lengths in the least amount of time possible, beginning on the right side of the semicircle.

Analysis and Interpretation of Data

This is a timed test, with better scores equating to lower times.

Statistics

■ Significant mean differences ($F = 5.39$, $p = 0.002$) for male athletes were shown for this test. Football back and line athletes and baseball players performed the test significantly faster than male gymnasts. Among women, basketball players and volleyball players performed the test significantly faster ($F = 7.54$, $p = 0.001$) than gymnasts (Lephart et al. 1991).

■ A strong correlation could not be found between the co-contraction test and any lower extremity physical characteristics such as hamstring and quadriceps peak torque values, torque acceleration energy, joint laxity, thigh circumference, and knee range of motion (Lephart et al. 1992).

■ No significant difference was found in co-contraction test scores between clients who had a patellar tendon autograft and those with an allograft anterior cruciate ligament (ACL) reconstruction (Lephart et al. 1993).

STEP-DOWN

▶ **Purpose:** To assess the client's ability to perform a coordinated and controlled eccentric and static stabilization movement.

▶ **Equipment:** 19 cm step, force plate platform.

Procedure

1. Position the client so that he or she has both feet on the step and is facing the force plate.

2. Instruct the client to place hands on hips and to take a single drop-step forward, leading with the test leg, and land on the force plate with the test leg only.

3. The client should be instructed that the lead foot should hit the force plate at the same time the trailing foot leaves contact with the step.

Analysis and Interpretation of Data

■ Multiple forms of data can be reviewed, for example center of pressure data including anterior and posterior displacement and medial-lateral displacement.

■ Stabilization time can also be assessed.

Statistics

■ For reliability statistics regarding this test, see table 9.1 on page 167.

■ Colby and colleagues (1999) found, when comparing step-down test scores of ACL reconstruction patients, that they had significantly longer stabilization times on the involved as compared to the uninvolved limb (1527 ± 216 ms and 892 ± 498 ms), respectively. These findings may be the result of an altered pattern of motion in compensation for increased knee instability (Colby et al. 1999; Gauffin & Troop 1992).

■ A significant decrease in peak vertical force was found in the injured limb of the ACL reconstruction group compared to the uninjured limb during the step-down test ($P < .01$). This may simply reflect an attempt to decrease vertical forces so that anterior tibial acceleration can be decreased through a different landing technique. This compensation pattern has been noticed by others as a flexed knee landing pattern (Gauffin & Troop 1992; McNair & Marshall 1994).

CARIOCA DRILL OR TEST

▶ **Purpose:** To assess the client's ability to move laterally while crossing the legs in a coordinated fashion as quickly as possible.

▶ **Equipment:** Floor marked with 12 m distance between two pieces of tape.

Procedure

1. Instruct the client to move laterally with a crossover step across two lengths of the 12 m distance.

2. Have the client begin by moving from left to right and then reverse direction following the first 12 m length, thus performing the test by moving a total of 24 m in the minimum amount of time possible.

Analysis and Interpretation of Data

This is a timed test, with better scores equating to lower times.

Statistics

■ Lephart and colleagues (1992) were unable to find a strong correlation between the carioca test and any lower extremity physical characteristics. These authors also found that scores for the carioca test were significantly different between subjects who could return to competition and those who could not. Those who could return to competition were able to perform the carioca test in 8.45 ± 1.93 s as compared to 17.31 ± 14.33 s for those who could not. Furthermore, Lephart and colleagues (1993) found no significant differences in the carioca test score between those who had undergone a patellar tendon autograft and those who had an allograft ACL reconstruction.

■ Quadriceps and hamstring peak torque as tested isokinetically and their correlations with this test are listed in table 9.2 on page 167.

LUNGE TEST

► **Purpose:** The lunge is a very common exercise in sport and orthopedic rehabilitation; therefore it seems reasonable to use it as a functional test. Performance of the lunge reflects lower extremity strength, balance, and flexibility (Foran 2000; Gray 2001). The lunge has been described in both the anterior and lateral directions.

► **Equipment:** Tape measure and standard firm flooring or other surface.

Procedure (Crill et al. 2004)

For the forward lunge:

1. Instruct the client to step forward with the dominant leg first, leaving the trailing leg outstretched behind.
2. Inform the client that the lunge leg should flex at the knee directly above the heel of the same leg; the feet should stay parallel to the direction of the lunge.
3. For scoring, measure the distance between the toe of the stance leg and the heel of the lunge leg.

For the lateral lunge:

1. Instruct the client to step out to the lateral side, leading with the dominant leg.
2. Inform the client that the lunge foot remains perpendicular to the direction of the lunge; the stance leg is to remain extended, with the stepping leg flexed slightly at the knee.
3. For scoring, measure from the medial aspect of the stance heel to the medial aspect of the lunge heel.

Analysis and Interpretation of Data

■ The major determination of objective data comes from lunge distance. This test provides a fair amount of subjective information that can be gleaned also. A successful lunge requires the client to be able to lunge and return to the starting position without loss of balance. The anterior lead foot should remain in its position upon landing. This should occur without rotation out of position. Movements of supination and pronation are allowed as long as the lead foot remains fixed. An attempt is considered failed if the client loses balance enough to have to move the foot or if he or she has to stabilize by touching the ground with hands or other body parts. Addition-

ally, excessive movements of the lead leg into hip adduction and internal rotation may be indicative of hip weakness. Other subtle differences may be seen from typically inflexible muscles such as the gastrocnemius-soleus complex, hamstring, and gluteus maximus.

Forward lunge.

Lateral lunge.

(continued)

(continued from previous page)

■ Another form of the lunge test is performed onto a force plate; movement characteristics can be measured in the resting standing position, in the lunge forward with one leg, and during the return to the starting position (Mattacola et al. 2004). Variables such as distance, impact index, and force impulse have been measured. It is assumed that the larger the distance the person can lunge, the better he or she can control the lower extremity. Additionally, the larger the lunge, the greater the force production at impact. A longer contact time would indicate a slower movement, which may be interpreted as decreased function. The impact index, expressed as a percentage of the client's body weight, quantifies the maximal vertical force exerted by the stepping onto the force plate during the lunge.

Statistics

■ Mattacola and colleagues (2004) assessed a group of ACL reconstruction patients (n = 18) who had undergone a bone–patellar tendon–bone or quadruple semitendinosus–gracilis tendon procedure and a group of control subjects (n = 18) using the forward lunge onto a force plate. The control group demonstrated greater measures of functional ability than the ACL reconstruction group. No significant differences were found between groups for lunge distance or contact time; significant differences were noted for impact index and force impulse. A significant main effect was noted for extremity ($F[1,34] = 7.300$, $[P = .006]$) as well as a significant group × extremity interaction ($F[1,34] = 8.541$, $[P = .006]$) for the impact index. The impact index was greater for the uninvolved leg (29.4% body weight) than the involved leg (23.7% body weight) during foot strike for the ACL reconstruction individuals. The clinical relevance of this is that it may indicate decreased eccentric control for the ACL-reconstructed leg. The force impulse was also greater for the uninvolved leg (88.6% body weight/s) than for the involved leg in the ACL reconstruction group (82.1% body weight/s), indicating that the ACL-reconstructed leg performed less work.

■ Similarly, Alkjaer and colleagues (2002) evaluated the forward lunge in two ACL-deficient patient groups and controls. The lunge was performed more slowly by noncopers than by the controls or the coper group (1.27 s, 1.08 s, and 0.99 s, respectively). As mentioned previously, a decreased loading time leads to a decreased amount of time during which the knee is actually loaded. An increased loading time indicates an inability to control the knee dynamically as it is loaded, which can be interpreted as decreased performance.

KNEE BENDING IN 30 SECONDS

▶ **Purpose:** To assess the client's ability to perform repetitive knee bending on one leg, as well as the capability to perform fast changes between eccentric and concentric muscle force production over the knee joint.

▶ **Equipment:** Solid flooring with tape marking for starting position; it may be advisable to have a stable chair or plinth behind the client or something solid to grab onto in case of loss of balance.

Procedure (Roos et al. 2001; Bremander et al. 2007)

1. Position the client so that he or she is standing directly behind the tape marking on the flooring. The long axes of the feet are aligned with a straight line and the toes placed on a second, perpendicular line.
2. Support the client's fingertip.
3. Instruct the client to bend the test knee, without bending forward from the hip, until he or she can no longer see the line along the toes (about 30° of knee flexion).
4. Record the number of knee bends performed in 30 s.

Analysis and Interpretation of Data

No known statistical data has been reported for this test.

Statistics

Intraclass correlation coefficient (ICC) of test–retest reliability was 0.92 (Bremander et al. 2007) and was able to discriminate with regard to age, gender, and symptoms among clients postmeniscectomy or with radiographically verified knee osteoarthritis.

ONE-LEGGED CYCLIC HOP TEST

▶ **Purpose:** To assess the client's ability to repetitively hop or jump off lower extremities for time, height, contact time, frequency of hops or jumps, and other parameters.

▶ **Equipment:** Force plate, contact mat, or Jumpergometer (Fitronic, Bratislava).

Procedure

1. Instruct the client to jump as high as possible and to keep contact with the platform as brief as possible, as if it were a hot plate.

2. Instruct clients to jump (both legs) or hop (single leg) and to land on either both legs or one leg.

3. Instruct clients to keep both hands on the hips to eliminate their use in generating momentum.

Analysis and Interpretation of Data

Various parameters can be assessed. Height, contact time, and frequency can all be determined.

Statistics

■ Petschnig and colleagues (1998), who had subjects keep both hands on the hips, found ICC for between-day comparisons on the one-legged jump of 0.89.

■ In another version of the one-legged cyclic hop test, Pfeifer and Banzer (1999) had subjects perform 20 cycles of hopping with each leg. Subjects were instructed to keep the ground contact time as short as possible but still to hop as high as possible. Subjects were allowed to choose their own hopping frequency. In assessment of a group of 39 subjects with arthroscopically assisted ACL reconstruction and 20 control subjects, significant interactions were seen between the involved leg and a number of parameters in the experimental group. Interactions occurred for jumping time (185 ± 49 ms), range of motion (19.2 ± 8.7°), maximum knee angle (39.0 ± 9.7°), angle speed eccentric contraction (118 ± 54°/s), and angle speed concentric contraction (171 ± 63°/s).

SINGLE-LEG INCLINED SQUAT TEST (20 SECONDS AND 50 REPS)

▶ **Purpose:** The single-leg inclined squat test has been described by Munich and colleagues (1997). It is felt that the use of a weight-bearing functional test in this manner may be better indicated than other more stressful tests in the acute stages of injury and recovery. Munich and colleagues suggest that this test may be appropriate for clients with total knee arthroplasty, ACL reconstruction, and meniscal repairs, who usually are partial weight bearing during the first few days of rehabilitation.

▶ **Equipment:** As described by Munich and colleagues (1997), the Total Gym (Engineering Fitness International, Inc., San Diego, CA) is used to allow the client to lie in an inclined position.

Procedure

1. Use the uppermost notch on the Total Gym for maximum resistance.
2. Position the client so that the dominant foot is on the force plate, and record this position for future test replication.

3. Instruct the client to hold the other leg in the air.
4. Monitor that the hip and knee angle reaches 90° at the bottom position, and set a distance limiter so that the extremity will not go beyond this point.
5. For a 20 s test, instruct the client to squat up and down as fast as possible for 20 s, and record the number of completed squats as the score.
6. In another version of the test, ask the client to squat up and down as fast as possible to complete 50 repetitions, and record the time in seconds.

Analysis and Interpretation of Data

Side-to-side comparison is made to determine symmetry between lower extremities.

Statistics

Reliability for these two testing procedures has been studied, with each testing day separated by a full week. The ICC for the 50-repetition timed test was 0.80; the ICC for the 20 s repetition test was 0.89.

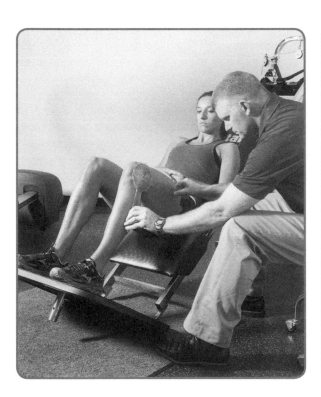

PROTOCOL FOR 1RM TESTING

This testing protocol for the 1RM (Kraemer & Fry 1995) is used in the following three tests: 1RM back squat, 1RM leg press, and 1RM bench press.

Procedure

1. Have the client warm up by doing 5 to 10 repetitions of the exercise at 40% to 60% of the estimated 1RM.

2. During a 1 min rest, have the client stretch the muscle group or groups being tested.

3. Have the client perform three to five repetitions of the exercise at 60% to 80% of the estimated 1RM.

4. Increase the weight conservatively, and have the client attempt the 1RM lift.

5. If the lift is successful, instruct the client to rest 3 to 5 min before attempting the next weight increment.

6. Follow this procedure until the client fails to complete the lift (the 1RM is typically achieved within three to five trials).

7. Record the 1RM value as the maximum weight lifted.

1RM BACK SQUAT

▶ **Purpose:** To determine 1RM back squat strength and trunk and lower extremity strength.

▶ **Equipment:** Standard squat rack, weight belt (preferable). In addition to the equipment, two to three spotters are needed.

Procedure

1. Instruct the client to grasp the bar with a closed, pronated grip (width depends on bar position).

2. Have the client step under the bar and position the feet parallel to each other.

3. Instruct the client to place the bar in a balanced position on the upper back and shoulders in one of two locations:

   - Low bar position: across posterior deltoids at the middle of the trapezius (using a hand grip wider than shoulder-width)

   - High bar position: above posterior deltoids at base of the neck (using a hand grip only slightly wider than shoulder-width)

Correct, front view.

Correct, side view.

Incorrect, front view.

Incorrect, side view.

(continued)

(continued from previous page)

4. Instruct the client to lift the elbows up to create a "shelf" for the bar using the upper back and shoulder muscles; chest is held up and out, and bilateral scapulae are adducted and depressed.

5. Both spotters assist with lifting the bar and gradually place it in one of the positions listed. Instruct the client to extend the hips and knees to lift the bar off the rack.

6. Instruct the client to take one or two steps backward and position the feet shoulder-width apart (or slightly wider) with toes pointing slightly outward.

7. Have the client allow the hips and knees to slowly flex while keeping the torso-to-floor angle relatively constant, back flat, chest up, and elbows up high. Do not allow flexing of the torso or rounding of the back.

8. Have the client continue to flex the hips and knees until the thighs are parallel to the floor as the spotters hold bilateral hands just below the bar on the respective sides.

9. Have the client extend the hips and knees to reach the initial starting position, with spotters in the same position as during the descent.

10. After the repetition is completed, have the client step forward and with the assistance of the spotters gradually lower the bar to the rack.

11. Repeat this sequence until the 1RM is determined according to the protocol (Kraemer & Fry 1995) described on page 140.

12. As to breathing technique, instruct the client to inhale deeply to maintain intrathoracic pressure and prevent bending forward, arching the back, or passing out.

13. With maximum testing, instruct the client to hold his or her breath prior to descending in the squat and to exhale when at or near the top of the squat. Instruct the client to inhale and exhale at the top of the squat between repetitions.

Analysis and Interpretation of Data

▨ Percentile and normative values for various populations are listed in tables 9.3 through 9.6, pages 168 through 170.

▨ A strong correlation has been shown between maximal strength in parallel squats and sprint performance and jumping height (Wisloff et al. 2004).

Statistics

ICC (r) = 0.93 (Stone & O'Bryant 1987); ICC (r) = 0.94 (Sewell & Lander 1991); ICC (r) = 0.99 (Giorgi et al. 1998); ICC (r) = 0.95 (Hickson et al. 1994); ICC (r) = 0.92 (Sanborn et al. 2000); ICC (r) = 0.99 (McBride et al. 2002).

1RM LEG PRESS

▶ **Purpose:** To determine 1RM lower extremity strength.

▶ **Equipment:** Standard leg press apparatus; in addition to the equipment, one or two spotters are needed.

Procedure

1. Instruct the client to position him- or herself properly on leg press apparatus (bilateral legs shoulder-width apart, or slightly more, and bilateral feet turned slightly out).

2. Instruct the client to extend bilateral legs (with assistance of one or two spotters) without locking out the knees.

3. Have the client gradually flex bilateral hips, knees, and ankles until he or she achieves a 90° angle at the knee (unless a predetermined angle other than 90° is to be assessed).

4. Have him or her then extend bilateral legs until bilateral knees are straight without hyperextending or locking out bilateral knees.

5. Repeat this sequence until you have determined the 1RM via the protocol (Kraemer & Fry 1995) described on page 140.

Analysis and Interpretation of Data

Age and gender normative values for the 1RM leg press are listed in table 9.7 on page 170.

Statistics

ICC (r) = 0.89 (Hoeger et al. 1990); ICC (r) = 0.99 (Kraemer et al. 2000); ICC (r) = 0.99 (Rhea et al. 2002).

Note

Single-leg press testing can be implemented similarly, with comparison of side to side to assess limb symmetry.

1RM BENCH PRESS

▶ **Purpose:** To determine the client's 1RM strength in the bench press maneuver.

▶ **Equipment:** Standard bench press apparatus; in addition to the equipment, one or two spotters are needed.

Procedure

1. Ask the client to grip the bar with the hands shoulder-width and the thumbs touching the outside edge of the shoulder.

2. In most cases, have the client do a warm-up set.

3. Have the client lift the weight bar off the rack with the assistance of the spotters and lower it to his or her chest.

4. After the bar has been stable for at least 1 s, give the command to "press," at which time the client presses the weight bar into full extension.

5. Allow a rest period of at least 3 min for recovery between lifts.

6. Repeat this sequence until you have determined the 1RM via the protocol (Kraemer & Fry 1995) described on page 140.

Analysis and Interpretation of Data

Percentile and normative values are given for various populations in tables 9.8 through 9.12, pages 171 through 174.

Notes

■ The risk of injury and the length of time taken to determine a 1RM (especially with a group of individuals) can be potential drawbacks to 1RM testing. Although 1RM tests can be safely administered to clients of all ages, sometimes it is preferable to estimate the 1RM. Submaximal testing has been used to avoid the problems just mentioned. The use of submaximal loads to predict maximal strength has been shown to have correlation coefficients >0.90 (Lander 1984; Mayhew et al. 1992). The number of repetitions performed at selected percentages of the 1RM varies among the different exercises, as well as within an exercise (Hoeger et al. 1987, 1990). If more than 10 repetitions are performed, the National Football League 225 lb bench press test loses its validity and underestimates actual strength (Ware et al. 1995). Equations have been developed for prediction of 1RM strength with submaximal testing:

- (0.033) (Reps) (Weight lifted) + Weight lifted (Epley 1985).
- (100) (Weight lifted) / [101.3 − 2.67123 (Reps)] (Landers 1985).
- Weight lifted (pounds) / [1.0278 − 0.0278 (Reps)] (Brzycki 1993).

■ Kuramoto and Payne (1995) have developed prediction equations to estimate a 1RM from a submaximal endurance test in middle-aged and older women. The client completes as many repetitions as possible using a weight equivalent to 45% of her body mass. To estimate 1RM, use the following equations:

- Middle-aged women (40-59 years) ($r = 0.94$):

  1RM = [1.06 × Weight lifted (kg)] + (0.58 × Reps) − (0.20 × Age) − 3.41.

- Older women (60-70 years) ($r = 0.90$):

  1RM = [0.92 × Weight lifted (kg)] + (0.79 × Reps) − 3.73.

■ Another way of ascertaining 1RM is with a predetermined relationship shown between 1RM and number of repetitions allowed during testing.

% 1RM	Number of repetitions allowed
100	1
95	2
93	3
90	4
87	5
85	6
83	7
80	8
77	9
75	10
70	11
67	12
65	15

Data from Baechle and Earle, 1989; Brzycki 1993; Chapman et al. 1998; Epley 1985; Lander 1984; Mayhew et al. 1992; Morales & Sobonya 1996; Wathen 1994.

SINGLE-LEG SQUAT

► **Purpose:** To assess lower extremity functional strength, neuromuscular control, and dynamic flexibility via a single leg. With monitoring from the front, assessment can be the amount of frontal plane movement. Is it excessive, for example? With monitoring from the side, assessment can include measurement of degree of hip, knee, or ankle flexion with a standard goniometer. Assessment can also include the extent of sagittal (front to back) plane movement (specifically at the trunk).

► **Equipment:** Stable testing surface.

Procedure (Clark 2000)

1. Have the client stand in single-leg stance on the leg to be assessed.
2. Preferably, ask the client to perform the test without shoes on so that you can monitor foot biomechanics.
3. Instruct the client to keep the trunk and upper body as upright as possible during the squat.
4. Instruct the client to place bilateral hands on hips.
5. Instruct the client to squat down as far as possible toward the floor, bending at the hip, knee, and ankle as far as necessary.

Analysis and Interpretation of Data

No objective data have been reported for this test. Assessment should include measuring the amount of knee flexion obtained while squatting, although subjective evaluation of technique is very important. Testers should note loss of balance or uncontrolled motions at the foot, knee, or hip that may be caused by proximal or distal muscle weaknesses.

Statistics

No studies with statistical data were found.

Frontal plane, no compensation.   **Compensation in frontal plane.**   **Sagittal plane, no compensation.**   **Compensation in sagittal plane.**

ALTERNATE SINGLE-LEG SQUAT TEST

▶ **Purpose:** To determine the client's 1RM strength in the single-leg squat maneuver.

▶ **Equipment:** Barbell or dumbbells, resistance band, squat rack.

Procedure (McCurdy & Langford 2005)

1. Instruct the client to position him- or herself in the squat rack by placing the metatarsophalangeal area of the foot of the nontest leg on the posterior support in order to isolate the use of the lead leg.

2. Prior to testing, have the client perform a squat to a 90° angle.

3. To mark squat depth, you can stretch a resistance band across the frame; set the band at a height that allows the client's hamstrings to touch the band to attain the 90° angle at the hip and knee.

4. For proper starting position, instruct the client to center the lead leg in the squat rack approximately 1 in. (2.5 cm) in front of the measurement band, with the leg and upper body in normal anatomical stance.

5. Instruct the client to flex the knee of the test leg at 90° with the hip slightly hyperextended to place the top of the foot on the pad of the posterior bar.

6. After a warm-up with light weight, increase the weight and have the client complete 5 to 10 repetitions followed by a 3 to 5 min rest between each successive set.

7. After you have increased the weight 20% to 30%, have the client attempt a 1RM.

8. If the attempt is successful, increase the weight 10% to 20%.

9. If the attempt is unsuccessful, have the client attempt one final trial after 5% to 10% of the weight has been removed.

Analysis and Interpretation of Data

▪ For a successful trial, the client's hamstring has to touch the resistance band.

▪ Testing includes both dominant and nondominant lower extremities.

▪ This test requires some subjective observation to assess for excessive shifting or leaning of the trunk or lead leg.

▪ Normative values for the single-leg squat in both men and women and dominant and nondominant legs are listed in table 9.13 on page 174.

Statistics

▪ McCurdy and Langford (2005) studied the modified single-leg squat in healthy men and women. They found no significant difference between subjects' dominant and nondominant leg strength.

▪ McCurdy and colleagues (2004) found acceptable reliability for the IRM single-leg squat in trained men ($r = 0.98$), untrained men ($r = 0.99$), trained women ($r = 0.99$), and untrained women ($r = 0.97$).

STANDING LONG JUMP

▶ **Purpose:** To assess lower extremity functional strength, neuromuscular control, and dynamic power.

▶ **Equipment:** Standard tape measure or testing mat to measure horizontal distance jumped, flat nonslip testing surface.

Procedure

1. Position the client so that he or she is standing with bilateral feet a comfortable distance apart just behind the starting line.
2. You can choose from various hand positions (hands on hips or behind back, or free for use during the jump); but keep the hand placement consistent.

3. After proper warm-up, have the client perform a maximal jump along the testing surface.
4. Allow the client to perform a countermovement prior to jumping.
5. Instruct the client to hold the landing to allow the distance to be marked.
6. Note that for the jump to qualify as a legitimate jump, the client must land without losing balance, falling, or taking an additional step.
7. Measure the distance from the starting line to the back of the heel closest to the starting line.
8. Have the client perform three trials, with adequate rest for recovery between trials.
9. Record the average of the three trials.

Analysis and Interpretation of Data

■ The contributions to the propulsive phase of this test have been determined to be hip 45.9%, knee 3.9%, and ankle 50.2% (Robertson & Fleming 1987).
■ Various normative values for the standing long jump test are listed in tables 9.14 through 9.16.
■ A strong correlation ($r = 0.805$; n = 32) has been found between 1RM squat and standing broad jump performance; maximum strength is also strongly related to jumping ability (Koch et al. 2003).

Statistics

■ ICC = 0.95; coefficient of variation (CV) = 2.4% (Markovic et al. 2004).
■ ICC = 0.98, using a 15 min interval between jump tests in which clients sat quietly without engaging in overt activity (Koch et al. 2003).
■ Test–retest reliability has been shown to vary from 0.91 to 0.96 (Manske et al. 2003; Unger & Wooden 2000; Markovic et al. 2004).

Note

A version of this test includes allowing the client to freely swing bilateral arms rather than keeping them placed on the hips. Please refer to the vertical jump description for detail on contributions of the upper extremities during jumping.

SINGLE-LEG HOP FOR DISTANCE

▶ **Purpose:** The single-leg hop for distance test was initially described by Daniel and colleagues (1982) for assessing knee function. It is a very simple test to perform that can be done in an area no more than 15 ft (4.6 m) in length (Daniel et al. 1982). This functional knee test is included in the International Knee Documentation Committee (IKDC) knee test protocol (Andersen 1994; Daniel & Andersen 1991). The test can also be utilized to assess overall single lower extremity power for side-to-side comparison.

▶ **Equipment:** Standard tape measure or testing mat (SBP, Toronto, Canada) to measure horizontal distance hopped, placed on a flat nonslip testing surface.

Procedure

1. Ask the client to stand with the front of his or her shoes behind the starting line.

2. You can choose from various hand positions (hands on hips or behind back, or free for use during the jump), but keep the hand placement consistent.

3. Ask the client to hop horizontally as far as possible, landing with the knee bent to help decrease the risk for knee injury.

4. Use the spot that marks placement of the posterior heel to measure hop distance.

Analysis and Interpretation of Data

■ Note that for the hop to qualify as legitimate, the client must land without losing balance, falling, or taking an additional step.

■ The landing position must be held for at least 2 s. In most cases quantitative assessment using the single-leg hop test is done via comparisons between the involved and uninvolved leg and has been shown to be useful in evaluating functional limitations (Petschnig et al. 1998).

■ It has also been shown that the uninvolved leg following ACL reconstruction, regardless of limb dominance, can be used as a control (Petschnig et al. 1998).

■ The distance hopped by healthy men who served as controls did not differ from the distance hopped on the uninvolved leg in patients at 13 weeks or 54 weeks following ACL reconstruction. This would suggest that the uninvolved leg may be used as a

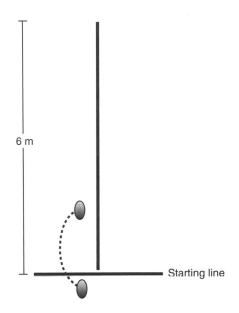

reference in hop tests for distance (Petschnig et al. 1998).

■ If the uninvolved leg has a history of injury or surgery, it can be used as a reference only with caution (Elliot 1978).

Statistics

■ According to Itoh and colleagues (1998), 42% of ACL-deficient patients exhibited abnormal lower limb symmetry when performing the single-leg hop. They reported this percentage as the test's "sensitivity," expressed as the percentage of ACL-deficient patients who showed abnormal lower limb symmetry values in a given test. Both Tegner and colleagues (1986) and Noyes and colleagues (1991) reported a low sensitivity of the single-leg hop, 38% and 58%, respectively. Petschnig and colleagues (1998), assessing 55 ACL reconstruction patients, reported a specificity of 98% and a false positive rate of only 2% with the single-leg hop test. Noyes and colleagues (1991) reported that the combination of the single-leg hop and the timed hop test can increase the sensitivity to 62% if the tested knee is rated as abnormal when performance in at least one of the two tests is found to be beyond the normal range. Noyes and colleagues (1991) recommended conducting at least two functional tests to discriminate subtle functional abnormality. Itoh and colleagues (1998), when combining the figure 8 hop

(continued)

(continued from previous page)

test, the up–down test, and the side-hop test with the single-leg hop test, increased sensitivity to 82%, reflecting the true functional disabilities of the ACL-deficient patient. DeCarlo and Sell (1997) concede this important finding and also report that the results of the single-leg hop test should not be used alone but rather in conjunction with other objective and subjective data gathered during the clinical visit.

▪ More recent reliability analyses have shown ICC = 0.93 (Bremander et al. 2007); ICC = 0.92 with a correlation (Pearson r) of 0.37 with the Lower Extremity Functional Scale and 0.48 with the Global Rating Scales (Reid et al. 2007).

▪ Petschnig and colleagues (1998) reported that the single-leg hop appears to be a valuable screening assessment especially when used to determine limb symmetry index (LSI). To determine LSI, the mean of the involved limb is divided by the mean of the uninvolved limb, and the result is multiplied by 100. In those tested 13 weeks following ACL reconstruction, 93% of the patients had an abnormal LSI, while at 52 weeks only 28% still had an abnormal score. Thus the single-leg hop test returned to normal in approximately 72% of patients between 13 and 52 weeks. Barber and colleagues (1990) reported that the percentage of the normal population that exhibits an LSI of 90% is approximately 81%; the percentage that exhibits an LSI of 85% is 93%; and the percentage of the normal population that reaches an LSI of 80% will be 100%.

▪ Engstrom and colleagues (1993) assessed patients' single-leg hop scores following conservative treatment for an ACL injury and determined that there were significant differences between the injured and uninjured sides ($p < 0.001$). There was, however, no significant correlation between one-leg hop ratio and indices of joint laxity or results for the pivot shift test. There was a significant correlation between knee extensor peak torque tested by isokinetic dynamometry and the single-leg hop for the injured ($p < 0.001$) and the uninjured side ($p < 0.001$). The authors suggest that the single-leg hop could be used as an easy field test reflecting knee extensor strength.

▪ Barber and colleagues (1990) found a statistically significant relationship between abnormal LSI in the single-leg hop test and subjective limitations in

sprinting and jumping/landing ($p < 0.01$). Additionally they found a statistically significant relationship between a 60°/s quadriceps percentage deficit score and abnormal LSI in the one-legged hop for distance test ($p < 0.01$). Eighteen of 27 patients had abnormal scores on their isokinetic test. Twelve of those 18 had an abnormal LSI in the single-leg hop for distance test.

▪ Sekiya and colleagues (1998) reported low correlation coefficients between hop index and muscle strength index, suggesting that the single-leg hop test is influenced by many factors such as ankle, hip, trunk, and upper limb muscle strength; proprioception; and rotatory instability.

▪ Normative values and reliability of the single-leg hop for distance test among various populations are listed in table 9.17 on page 176.

▪ Bandy and colleagues (1994) have concluded that the functional hop tests are reliable and may provide objective assessment of progress in a rehabilitation program as well as the ability of the extremity to tolerate external forces.

▪ Colby and colleagues (1999) used a slight variation of the single-leg hop test in which subjects jumped onto a force plate platform. The authors found that the mean standard deviation of vertical force was significantly greater for the injured limb (injured limb = 194 ± 56 N; uninjured limb = 179 ± 81 N) in the ACL-deficient groups, while the mean standard deviation of vertical force was significantly less for the injured limb (injured limb = 170 ± 41 N; uninjured = 186 ± 27 N) in the ACL reconstruction groups ($p = 0.05$).

▪ Reliability and correlations of the single-leg hop to various parameters are listed in tables 9.18 and 9.19 on pages 178 and 179.

Note

Single-leg hop performance was impaired in the injured limb of ACL-deficient patients compared to their non-involved limb ($p < 0.05$) (Gauffin et al. 1990). This impaired performance in the involved limb was not necessarily correlated to a reduction in muscle strength. Gauffin and coworkers also found a difference ($p < 0.001$) between hop performance on the noninjured leg of these patients and performance of a reference group of normal noninjured subjects.

VERTICAL JUMP TEST

► **Purpose:** To assess the client's explosive power in a vertical direction.

► **Equipment:** Vertec (or, if Vertec is not available, a ladder, wall with high ceiling, a landing area with flat nonslip surface, chalk).

Procedure (Seminick 1994; Chu 1996)

1. Position the client so that he or she is standing with equal weight on bilateral lower extremities, which are approximately shoulder-width apart.

2. Have the client push back the highest reachable marker on the Vertec; this is the zero starting position. (If you are not using the Vertec or similar vertical jump measurement device, have the client hold a piece of chalk in the hand closest to the wall on the side of measurement, and instruct him or her to reach as high as possible and make a mark on the wall; this is the zero starting position.)

3. Record the mark as the zero starting position.

4. For the jump, you can choose from various opposite-hand positions (hand on hip or behind back, or free for use during the jump), but keep the hand placement consistent.

5. Instruct the client not to move the feet and to flex at the knee, hip, and ankle; jump; and push back the highest reachable marker on the Vertec. (If not using the Vertec, instruct the client not to move the feet and to flex at the hip, knee, and ankle; jump; and place a second chalk mark as high as possible on the wall.)

6. Note that reasons for disqualification during testing include any irregularity in placing the first mark on the wall (e.g., not having the feet flat) and taking a step prior to jumping.

7. If you are using the Vertec, the client's score is the vertical distance between the zero starting position and the highest reachable marker. (If not using the Vertec, the client's score is the vertical distance between the two chalk marks.) Record the best of three trials to the nearest 0.5 in. (1.25 cm).

Analysis and Interpretation of Data

■ Use of arm swing in bilateral upper extremities and use of a countermovement (squatting prior to jump) contribute to a significant increase in the vertical jump distance (Harman et al. 1990; Luthanen & Komi 1978; Shetty & Etnyre 1989).

■ The power the client can produce in this test can be derived from the following formula (Fox & Matthews 1974):

Power (watts) = 21.67 × Mass (kg) × Vertical displacement (m) × 0.5.

Zero starting position.

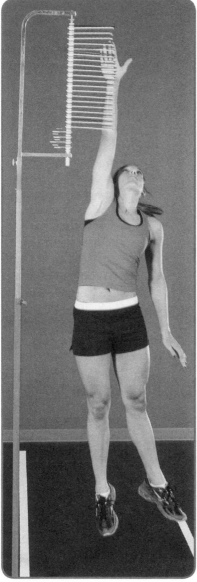

Maximum height of the vertical jump.

(continued)

(continued from previous page)

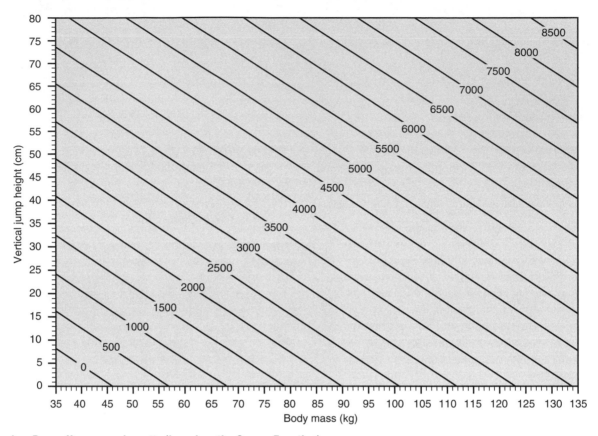

Leg Power Nomogram in watts (based on the Sayers Equation).

Reprinted from Kier et al. 2003 with permission from NSCA/Allen Press.

▦ For the propulsive phase of the standing vertical jump (as noted from jumping from a force platform), the contributions of the lower extremity musculature were hip 40%, knee 24.2%, and ankle 35.8% (Robertson & Fleming 1987).

▦ Percentile ranks and normative data for vertical jump heights in various populations are listed in tables 9.20 through 9.27 on pages 180 through 186.

▦ The Kier Peak Leg Power Nomogram provides a measure of leg power in the vertical jump using height jumped (cm) and body mass (kg) as input variables. The intersection of a horizontal line drawn from the jump height in centimeters (y-axis) with a vertical line drawn from the client's mass in kilograms (x-axis) provides the peak leg power in watts. Inclined lines indicate leg power in 100 W increments.

Statistics

▦ Test–retest reliability has been reported to range between 0.93 and 0.99 (Considine & Sullivan 1973; Glencross 1966).

▦ Reliability: ICC .96; CV = 3.0% (Markovic et al. 2004).

▦ ICC values ranged from 0.99 to 1.00 (Pauole et al. 2000).

▦ An r as high as 0.93 has been reported (Johnson 1986).

▦ A validity of 0.78 with the criterion of a sum of four track and field event scores has been reported (Johnson 1986).

▦ Markovic and colleagues (2004) concluded that the countermovement jump is the most reliable and valid field test for estimation of explosive power of the lower limbs in physically active college-age men.

▦ The use of arm swing and free-leg drive (single-leg takeoffs) has been shown to significantly influence vertical jump performance; therefore, vertical jump tests are not valid for assessment of leg extensor muscle function (Young et al. 2001).

Note

This test can also be performed with a single leg and comparison made from side to side for LSI.

TRIPLE JUMP FOR DISTANCE TEST

▶ **Purpose:** To assess the ability to perform a repetitive, explosive jumping movement off of bilateral lower extremities for assessment of repetitive power capacity.

▶ **Equipment:** Standard tape measure or testing mat to measure horizontal distance jumped, flat nonslip testing surface.

Procedure

1. Position the client so that he or she is standing with bilateral feet a comfortable distance apart just behind the starting line.

2. Instruct the client to bend and lift one leg at the knee in order to initiate the movement with a hop off the single leg.

3. Allow the client to swing bilateral arms as necessary (Markovic et al. 2004). Alternate hand placement of hands on hips, hands behind back, or hands free to move can be used, but it must be ensured that one method is used consistently.

4. After proper warm-up, have the client perform a maximal triple jump movement (hop onto starting leg, step onto opposite leg, and finish with a bilateral jump movement) along the testing surface.

5. Allow the client to perform a countermovement prior to jumping, and ask him or her to hold the landing to allow the distance to be marked.

6. Measure the distance from the starting line to the back of the heel closest to the starting line.

7. Have the client perform three trials with adequate rest for recovery between trials.

8. Record the average of the three trials.

Analysis and Interpretation of Data

■ The client can perform two series of jumps, each initiated with a different leg. The clinician can then compare these results and also compare bilaterally for LSI.

■ To determine LSI, divide the mean distance for the involved leg by the mean distance for the uninvolved leg and multiply the result by 100.

Statistics

ICC = 0.93, CV = 2.9% (Markovic et al. 2004); ICC = 0.97, SEM = 11.17 (Ross et al. 2002); ICC = 0.95, SEM = 17.1 (Bolgla & Keskula 1997); ICC = 0.94 (Bandy et al. 1994).

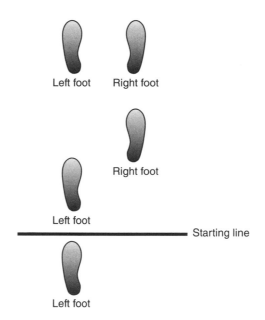

SIX-METER TIMED HOP

▶ **Purpose:** To assess the ability to produce power, speed, balance, and single lower extremity control over a specified distance, with specific emphasis on time for the single lower extremity. Comparisons can be made from side to side for LSI.

▶ **Equipment:** Tape to mark starting and stopping locations, low-pile rubber-backed carpet glued onto concrete floor or rubber-type mat flooring (stable, nonslip testing surface), stopwatch.

Procedure (Barber et al. 1990; Brosky et al. 1999)

1. Place two pieces of athletic tape on the floor, 6 m apart, to indicate starting and stopping locations.

2. Position the client so that he or she is standing behind the starting-line tape.

3. You can choose from various hand positions (hands on hips or behind back, or hands free to move), but use the chosen method consistently.

4. Instruct the client to perform a single-leg hop over the 6 m distance.

5. Encourage large, forceful one-legged propulsive hopping over the distance.

6. End the test when the back of the client's heel crosses the finish line.

7. Measure the time to the nearest 1/100 s.

8. To determine LSI, divide the mean time for the involved leg by the mean time for the uninvolved leg and multiply the result by 100.

Analysis and Interpretation of Data

■ This is a timed test, with better scores equating to lower times.

■ Clients must have confidence in the ability of their extremity to tolerate not only exploding as quickly as possible, but also landing on the single leg with a large eccentric component followed immediately by another hopping motion. It appears intuitive that a client with high levels of physical characteristics (strength, speed, power, agility, etc.) will have faster times than those with lower levels.

■ Wilk and colleagues (1994) reported that the single-leg timed 6 m hop test is one of the two functional tests with the most sensitivity and one of the best indicators of physical function.

Statistics

Test–retest reliabilities from studies involving various subjects and testing conditions for the 6 m timed hop test are listed in table 9.28 on page 186.

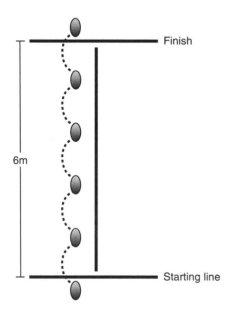

TRIPLE HOP FOR DISTANCE

▶ **Purpose:** To assess the ability to produce power, speed, balance, and single lower extremity control over a specified distance, with specific emphasis on distance covered by the single lower extremity.

▶ **Equipment:** Athletic or colored tape with a combined width of approximately 15 cm, placed parallel on gym floor; another alternative is a jump mat.

Procedure

1. Following a proper warm-up, position the client so that he or she is lined up directly behind the starting line, balancing on the lower extremity to be tested.

2. Instruct the client to perform three consecutive maximal-distance hops on one foot in a straight-line direction, that is, to stand on one leg and hop three times as far as possible with maximal effort.

3. Choose the hand position you prefer (hands on hips or behind back, or free to move), and instruct the client on that hand position.

4. Perform the test three times and average the values of the distance convered.

Analysis and Interpretation of Data

■ Side-to-side comparisons can be made for LSI.

■ Total distance covered can be measured for these comparisons, as well as for serial testing comparisons, specifically if one lower extremity is being rehabilitated.

■ As with all side-to-side comparisons, these types of assessments can be used to determine potential areas of concern for monitoring, the necessity of improvement to avoid imbalances from side to side, or both.

Statistics

■ Petschnig and colleagues (1998) found significant correlations with isokinetic strength scores of those following ACL reconstruction during testing at 13 weeks ($r = 0.48$) and 54 weeks ($r = 0.55$). Additionally, they found LSIs of 97.4 for the control groups, 73.0 for the group tested at 13 weeks, and 88.4 for the group tested at 54 weeks following reconstruction.

■ ICC for reliability 0.97 (Ross et al. 2002).

■ Recently Reid and colleagues (2007) obtained a reliability of $r = 0.88$ with this test, as well as correlation (Pearson r) values of 0.44 with the Global Rating of Change and 0.26 with the Lower Extremity Functional Scale, both of which are subjective ratings of function.

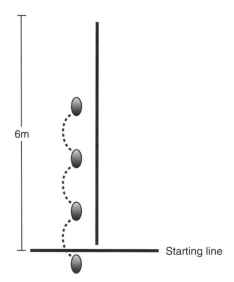

SINGLE-LEG CROSSOVER HOP FOR DISTANCE TEST

▶ **Purpose:** To assess the ability to produce power, speed, balance, and single lower extremity control over a specified distance, with specific emphasis on the distance covered by the single lower extremity and the ability to move laterally. Comparisons can be made from side to side to assess limb symmetry.

▶ **Equipment:** Athletic tape with a combined width of approximately 15 cm, placed parallel on gym floor; another alternative is a jump mat.

Procedure (Daniel et al. 1988; Barber et al. 1990)

1. Position the client so that he or she is standing on a single foot with the toe of that foot behind the starting line.

2. Instruct the client to perform three consecutive hops on one foot, crossing one of the marking strips with each hop.

3. Tell the client to hop as far as possible for all three hops and to hold the landing foot stationary on the last hop.

4. Do not allow the client to pause and compose him- or herself between hops.

5. Record as the score the distance from the starting point to the front of the client's foot on the final landing.

Analysis and Interpretation of Data

■ Hand position varies (hands on hips or behind back, or free to move); consistent hand placement should be used.

■ A test is considered failed if the client loses balance or the nontest leg makes contact with the floor.

■ To determine the LSI, divide the lowest value by the highest value and multiply the quotient by 100.

■ A modification of this test, using four hops rather than three, was developed by Andersen and Foreman (1996) and studied by Clark and colleagues (2002). It was felt that the test should impose an equal number of frontal plane (mediolateral) and rotational forces on the knee and provide more challenge.

■ Established normative values and reliability of the single-leg crossover hop for distance from various studies are listed in table 9.29 on page 187.

■ Wilk and colleagues (1994) found correlations between isokinetic strength testing and the crossover hop for distance. See table 9.30 on page 188.

Statistics

Statistical data are provided in table 9.29 on page 187 and table 9.30 on page 188.

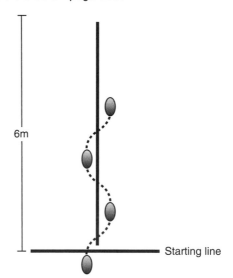

MAXIMAL CONTROLLED LEAP

▶ **Purpose:** Leaping is the projection of the body from one limb onto the other limb. A leaping test is a measure of force absorption in lower extremities, which may be a more accurate assessment of the client's ability to return to the previous level of function than a test in which the body is only propelled forward.

▶ **Equipment:** Stable testing surface with good traction; testing grid measured out with colored tape (or alternative), 1.91 cm wide and 3 m long and marked at 10 cm increments; tape measure or jump mat to measure distance jumped.

Procedure (Juris et al. 1997)

1. Position the client so that he or she is standing on a single lower extremity just behind the starting line as with the other hop tests described.

2. Have the client first perform a maximal single-leg horizontal hop as described for the single-leg hop. Position the client so that the non–weight-bearing leg at hip and knee angles approximating 90° of flexion and place their hands on their hips. These constraints are used to eliminate movement strategies involving leg and arm swing.

3. Three maximal hops are performed.

4. The client initiates the force absorption (maximal controlled leap) portion of the test in the same position as used in the single-leg hop. The client leaps (projects the body) from the leg initially placed on the floor as far as possible horizontally and lands on the opposite leg. The client is required to maintain their flexed hip-knee position throughout the takeoff phase, extending their leg only for landing. The primary requirement was for clients to "stick" the landing, arriving at a complete stop with their hands on their hips within 1 s.

Analysis and Interpretation of Data

■ The means of the three hops are used in the analysis.

■ The hop-and-stop test involved a maximal hop for distance and a maximal controlled leap.

■ Maximal hopping distance for each limb divided by the client's height (both in meters), $\times$ 100, normalizes the hopping distance to the client's stature.

■ Force production symmetry = the longer maximal hop value of the two limbs divided by that of the shorter hop $\times$ 100. If there is an involved lower extremity, the maximal value it achieves is divided by that of the uninvolved limb and the quotient is multiplied by 100. This provides a comparison of force production across limbs.

■ Stop-to-hop ratio = controlled leaping distance onto one limb divided by the maximal hopping distance of the opposite limb $\times$ 100. This ratio yields a measure of force absorption.

■ Stop symmetry = larger stop-to-hop ratio of the two limbs divided by the smaller ratio $\times$ 100. This provides a bilateral limb comparison of force absorption.

■ Lack of constraint (e.g., hands on hips) resulted in significantly longer ($p < 0.05$) hops compared to hops using constraints.

■ These tests of dynamic function seem to be less accurate without effective movement constraints.

■ Hop-and-stop test results for normal clients and patients, with criterion scores and sensitivity and specificity values for this test, are listed in table 9.32 on page 188.

Statistics

■ Data suggest that this protocol assesses functional and dysfunctional knees, and that force absorption may be more critical than force production in determining functional capacity (Juris et al. 1997).

■ No significant differences existed between limbs, raters, or the limb–rater interaction. The test, therefore, could easily and accurately be administered by different testers (Juris et al. 1997).

Note

A potential limitation stated by the authors (Juris et al. 1997) was that the clients were allowed an unlimited number of attempts to successfully complete the maximal controlled leap portion of the test.

STAIR HOPPLE TEST

▶ **Purpose:** To assess the ability to produce power, speed, balance, and single lower extremity control over a specified vertical and horizontal distance, with specific emphasis on speed over the distance covered by the single lower extremity. Comparisons can be made from side to side to assess LSI.

▶ **Equipment:** Stairs with 22 steps, standard stopwatch.

Procedure

1. Position the client so that he or she is standing on the test extremity with the other lower extremity held in hip and knee flexion so that it does not contact the flooring surface.
2. Ask the client to hop up and down 22 steps of a staircase (each step = 17.5 cm [7 in.] high) on first the dominant and then the nondominant leg.
3. Choose whether or not to allow for the client to use upper extremity contact with the wall or a banister for stability or other purposes.
4. Measure the time taken to perform the test in seconds.

Analysis and Interpretation of Data

Since this is a timed test, lower scores indicate better results.

■ Riseberg and Ekeland (1994) evaluated 35 ACL reconstruction clients who were a mean of 18 months postoperative to determine normative values.
■ Normative value for the noninvolved lower extremity was 22.7 ± 10.5 s and for the involved lower extremity was 25.2 ± 11.3 s.

Statistics

■ ICC = 0.94 for test–retest reliability (Goh & Boyle 1997).

Functional Ability Tests

The next four tests have been reported in the literature as a battery of tests of physical performance (Itoh et al. 1998). This battery includes the figure 8 hop test, which requires forced pivoting motions; the up–down test, which represents repetitive decelerations measuring hamstring reactive strength and anterior-posterior knee dynamic stability; the side-to-side hop test, which represents cutting and twisting motions and assesses rotatory lower extremity stability; and the single hop for distance, which measures leg power (Itoh et al. 1989, 1998). More than 95% of a control group exhibited symmetrical function in each part of the Functional Ability Tests (FAT), whereas in the ACL-deficient group, the percentage of patients who showed abnormal symmetry was 68% in the figure 8 hop, 58% in the up–down hop, 44% in the side hop, and 42% in the single hop (Itoh et al. 1998).

Tables 9.33 and 9.34 list normative values for the FAT, as well as comparisons between normal and ACL-deficient clients.

FIGURE 8 HOP TEST

▶ **Purpose:** To assess the ability to produce power, speed, balance, and single lower extremity control on a horizontal surface in multiple directions (specifically the figure 8 pattern), with specific emphasis on speed over the distance covered by the single lower extremity. Comparisons can be made from side to side to assess LSI.

▶ **Equipment:** Two standard stopwatches, colored or athletic tape.

Procedure

1. Position the client so that he or she is standing on the test lower extremity behind the starting line as with all the other hop tests described; the opposite leg is held in some hip and knee flexion to avoid foot contact with the floor.

2. Instruct the client to hop in a figure 8 pattern over a 5 m distance as fast as possible.

3. Choose the hand position you prefer (hands on hips or behind back, or free to use during hopping), but use the selected hand placement consistently.

4. Test the noninvolved lower extremity before the involved lower extremity.

5. Measure the time taken to perform two consecutive laps to the nearest tenth of a second.

Analysis and Interpretation of Data
See table 9.33 on page 189.

Statistics

▨ Intertester reliability has been shown to be 0.99 (Ortiz et al. 2005).

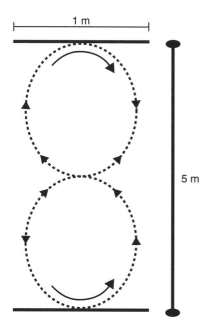

UP–DOWN TEST

▶ **Purpose:** To assess the ability to produce power, speed, balance, and single lower extremity control over a specified vertical and horizontal distance and number of repetitions, with specific emphasis on speed in a repetitive up-and-down manner for the single lower extremity. Comparisons can be made from side to side to assess LSI.

▶ **Equipment:** Two standard stopwatches, colored tape, a 20 cm high step.

Procedure

1. Position the client directly in front of the 20 cm high step, standing on the test lower extremity as previously described for other single lower extremity tests.

2. Require the client to hop vertically up and down the step for 10 repetitions as quickly as possible without turning backward.

3. Choose the hand position you prefer (hands on hips or behind back, or free to use during movement), but use the selected hand placement consistently.

4. Test the noninvolved leg before the involved leg.

5. Record the time taken to perform the 10 repetitions to the nearest tenth of a second.

Analysis and Interpretation of Data

See table 9.33 and table 9.34 on page 189.

Statistics

Interrater reliability of the up-and-down hop test has been shown to be 0.96 to 0.99 (Oritz et al. 2005).

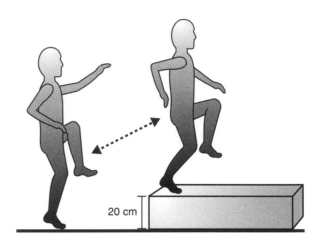

20 cm

SIDE-HOP TEST

▶ **Purpose:** To assess the ability to produce power, speed, balance, and single lower extremity control over a specified lateral and horizontal distance and number of repetitions, with specific emphasis on speed in a repetitive lateral manner for the single lower extremity. Comparisons can be made from side to side to assess LSI.

▶ **Equipment:** Two standard stopwatches, colored or athletic tape.

Procedure

1. Position the client so that he or she is standing next to the starting line with the nontest extremity held in some degree of hip and knee flexion to avoid ground contact.

2. Instruct the client to hop transversely or laterally more than 30 cm for 10 repetitions as quickly as possible.

3. Choose the hand position you prefer (hands on hips or behind back, or free to use during hopping), but use the selected hand placement consistently.

4. Test the noninvolved leg before the involved leg.

5. Record the time taken to perform the 10 repetitions to the nearest tenth of a second.

Analysis and Interpretation of Data

See table 9.33 and table 9.34 on page 189.

Statistics

Riseberg and Ekeland (1994) evaluated 35 ACL reconstruction patients who were a mean of 18 months postoperative to determine normative values for the side-hop test. Noninvolved lower extremity mean value for these patients was 13.4 ± 5.1 s and for the involved lower extremity was 13.9 ± 5.2 s.

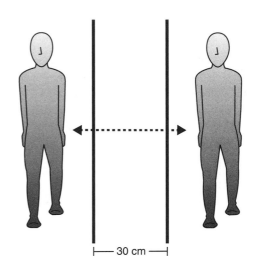

30 cm

SINGLE-HOP TEST

▶ **Purpose:** To assess the ability to produce power, speed, balance, and single lower extremity control over a specified horizontal distance and number of repetitions, with specific emphasis on speed in a repetitive lateral manner for the single lower extremity. Comparisons can be made from side to side to assess LSI.

▶ **Equipment:** Two standard stopwatches, colored or athletic tape.

Procedure

1. Instruct the client to hop forward as far as possible from the starting position marked with colored tape.

2. Choose the hand position you prefer (hands on hips or behind back, or free to use during hopping), but use the selected hand placement consistently.

3. Test the noninvolved leg before the involved leg.

4. Measure the maximum distance hopped from the starting-line tape to the client's heel.

Analysis and Interpretation of Data

See table 9.33 and table 9.34 on page 189.

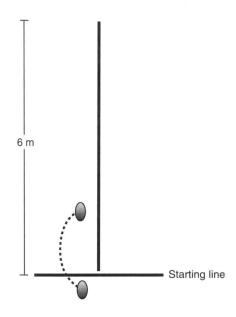

HEXAGON TEST (BILATERAL LOWER EXTREMITY JUMP)

▶ **Purpose:** To assess the ability to produce power, speed, balance, agility, and bilateral lower extremity and body control over a specified vertical and horizontal distance and over repetitions, with specific emphasis on speed in a repetitive up-and-down manner for bilateral lower extremities. Side-to-side comparisons are not possible as this is a jump and not a hop test. Readers are reminded that a hop is a single-leg maneuver and a jump is a double-leg maneuver.

▶ **Equipment:** Flat nonslip surface, a hexagon on the floor created with adhesive tape (color should contrast with that of the testing surface), stopwatch.

Procedure

1. Have the client start in the center of the hexagon and face forward throughout the test.

2. To begin the test, ask the client to perform a double-leg jump over the line immediately in front of him or her and jump back to the center.

3. Have the client work in a clockwise sequence until he or she has jumped over each of the six sides three times consecutively without stopping.

4. Have the client perform three trials, resting for 1 to 2 min after each of the first two.

5. Record the best time to the nearest tenth of a second.

6. You can choose another scoring option that is sometimes recommended for a more accurate assessment of the client's ability, which is to use the average of the three trials.

7. Note that disqualification results if the client lands on a line, loses balance, or has to take a step.

Analysis and Interpretation of Data

Bilateral hexagon test normative test values for athletes and college-aged females and males are listed in tables 9.35 through 9.37.

Statistics

ICC values range from 0.86 to 0.95 (Pauole et al. 2000).

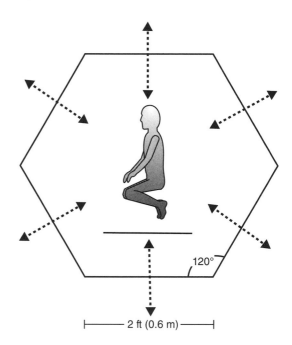

MODIFIED HEXAGON HOP TEST

▶ **Purpose:** To assess the client's ability to produce power, speed, balance, agility, and unilateral lower extremity and body control repetitively in a front-to-back as well as a lateral motion. Specific emphasis is placed on speed in the various directions for each lower extremity. Comparisons can be made from side to side to assess LSI.

▶ **Equipment:** Standard measuring tape to set up hexagon testing framework, two stopwatches, colored or athletic tape used to create a hexagon (2 ft [0.6 m] per side, 120° angles).

Procedure (Pauole et al. 2000; Ortiz et al. 2005)

1. Position the client inside the hexagon, standing on the test lower extremity with the other lower extremity in some degree of hip and knee flexion to avoid contact with the floor.

2. Have the client start in the center of the hexagon and face forward throughout the test.

3. Instruct the client to hop on the single lower extremity in and out of the hexagon, landing directly outside each side in consecutive order and not missing any sides.

4. Have the client work in a clockwise sequence until he or she has hopped over each of the six sides three times consecutively without stopping.

5. Allow the client to rest for 1 to 2 min prior to the next trial.

6. Have the client perform three trials, and record the best time to the nearest tenth of a second.

7. You can choose another scoring option, sometimes recommended for a more accurate assessment of the client's ability, which is to use the average of the three trials.

8. If the client does not cross all lines in completing one full lap, require him or her to rest an adequate amount of time and retest.

9. Note that disqualification results from landing on a line, losing balance, contacting the floor with the nontest lower extremity, or having to take a step.

Analysis and Interpretation of Data

■ Normal values (mean time ± SD) = 4.00 ± 0.44 (athletes; intertester testing) ($n = 23$).

■ Normal values (mean time ± SD) = 4.70 ± 0.82 (nonathletes; intertester testing) ($n = 24$).

■ Normal values (mean time ± SD) = 4.06 ± 0.45 (athletes; intratester testing).

■ Normal values (mean time ± SD) = 4.78 ± 0.82 (nonathletes; intratester testing).

Statistics

■ Intertester (ICC; SEM) = 0.95 (0.21) in athletes; 0.99 (0.20) in nonathletes.

■ Intratester (ICC; SEM) = 0.66 (0.47) in athletes; 0.76 (0.72) in nonathletes.

HOP TESTING AFTER FATIGUE

▶ **Purpose:** Since injuries often tend to occur at the end of a sporting event, when the participant is fatigued (Dugan & Frontera 2000; Feagin et al. 1987; Ostenberg & Roos 2000), a hop test assessment has been advocated for improved sensitivity in evaluation of lower extremity function after ACL reconstruction (Augustsson et al. 2004). Determining hop testing capacity after fatiguing the client may more accurately inform the clinician not only of the clients' power, but their potential to safely return to sport after ACL reconstruction.

▶ **Equipment:** Stopwatch, athletic tape.

Procedure

1. Position the client as with any other hop test, standing directly behind starting line with the nontest lower extremity held in some degree of hip and knee flexion to avoid ground contact.

2. Require the client to hop forward as far as possible from the starting position marked with colored tape.

3. Instruct the client to keep hands on hips throughout testing.

4. Measure the maximum distance hopped from the starting line to where client's heel landed.

5. Test the noninvolved leg before the involved leg.

Analysis and Interpretation of Data

■ Augustsson and colleagues (2004) regarded a hop as successful only when the client was able to keep his or her foot in place after landing (i.e., no extra hops for balance correction were allowed). Clients performed the test until they had done three successful hops with each leg.

■ The authors compared hop performance in two standardized test conditions, nonfatigued and fatigued. In the fatigued condition, clients performed the test immediately following preexhaustion exercise of the quadriceps muscle at 50% of 1RM. This fatigue protocol consisted of a single-joint quadriceps exercise (knee extension machine) until failure at 50% of 1RM strength.

■ In Augustsson and colleagues' (2004) testing procedure, clients performed the single-leg hop with both lower extremities, then performed the unilateral fatigue knee extension exercise until failure occurred at 50% of their respective 1RM, and then immediately performed the single-leg hops again. The leg tested first was randomized in each patient (Augustsson et al. 2004).

■ All of the clients in the Augustsson study (2004) were male ($n = 19$), with a mean time post ACL reconstruction of 11 months, and had exhibited normal single-leg hop symmetry values ($\geq 90\%$ compared with noninvolved lower extremity) as advocated by Juris and colleagues (1997). All clients were also tested for their 1RM on the leg extension machine.

■ Although no clients exhibited abnormal hop symmetry when nonfatigued, 68% showed abnormal hop symmetry for the fatigued test condition. Retesting 1RM in a fatigued state revealed that 63% of the clients exhibited 1RM strength scores below 90% of those for the noninvolved leg (Augustsson et al. 2004).

■ The number of hop trials following preexhaustion exercise of the quadriceps muscle at 50% of 1RM for the involved side was 5.0 (range of 3-9) and for the noninvolved side was 4.4 (range of 3-7), with no significant difference between sides ($p = 0.08$) (Augustsson et al. 2004).

Statistics

Test–retest reliability for the 1RM knee extension test: ICC = 0.96 for both sides (Augustsson et al. 2004).

TABLE 9.1

Step-Down Test Reliability

Parameter	Nondominant (SD)	Dominant (SD)
Anterior-posterior force	0.739 (0.743)	0.822 (0.764)
Medial-lateral force	0.872 (0.538)	0.557 (0.744)
Vertical force	0.896 (0.939)	0.933 (0.947)
Anterior-posterior center of pressure	0.828 (0.791)	0.866 (0.959)
Medial-lateral center of pressure	0.889 (0.856)	0.911 (0.850)

Colby et al. 1999

TABLE 9.2

Research Utilizing the Carioca Test

Authors	Subjects	Condition	Results
Lephart et al. 1992	41 subjects with unilateral anterior cruciate ligament insufficiency	Concentric quadriceps and hamstring peak torque, Q/H peak torque ratios at 60°/s and 270°/s	Correlation QPT 60°/s -0.30 HPT 60°/s -0.22 Q/H PT 60°/s -0.09 QPT 270°/s -0.27 HPT 270°/s -0.19
Keays et al. 2001	31 patients with unilateral anterior cruciate ligament reconstruction with hamstring autograft	Presurgery Postsurgery (6 months)	Outcomes 16.49 ± 4.24s 12.50 ± 2.37s

QPT = quad peak torque (isokinetic); HPT = hamstring peak torque; s = seconds.

TABLE 9.3

Percentile Values for the 1RM Squat in High School and College Football Players

% rank	HIGH SCHOOL (14-15 YEARS)		HIGH SCHOOL (16-18 YEARS)		NCAA DIVISION III		NCAA DIVISION I	
	lb	kg	lb	kg	lb	kg	lb	kg
90	385	175	465	211	470	214	500	227
80	344	156	425	193	425	193	455	207
70	325	148	405	184	405	184	430	195
60	305	139	365	166	385	175	405	184
50	295	134	335	152	365	166	395	180
40	275	125	315	143	365	166	375	170
30	255	116	295	134	335	152	355	161
20	236	107	275	125	315	143	330	150
10	205	93	250	114	283	129	300	136
$\bar{X}$	294	134	348	158	375	170	395	180
SD	73	33	88	40	75	34	77	35
n	170		249		588		1074	

Adapted, by permission, from J. Hoffman, 2006, *Norms for fitness, performance, and health* (Champaign, IL, Human Kinetics), 36-37.

TABLE 9.4

Percentile Values for the 1RM Squat in NCAA Division I Female Collegiate Athletes

% rank	BASKETBALL		SOFTBALL		SWIMMING		VOLLEYBALL	
	lb	kg	lb	kg	lb	kg	lb	kg
90	178	81	184	84	145	66	185	84
80	160	73	170	77	135	61	171	78
70	147	67	148	67	129	59	165	75
60	135	61	139	63	120	55	153	70
50	129	59	126	57	116	53	143	65
40	115	52	120	55	112	51	136	62
30	112	51	112	51	104	47	126	57
20	101	46	94	43	101	46	112	51
10	81	37	76	35	97	44	98	45
$\bar{X}$	130	59	130	59	118	54	144	65
SD	42	19	42	19	19	9	33	15
n	86		97		35		62	

Reprinted, by permission, from J. Hoffman, 2006, *Norms for fitness, performance, and health* (Champaign, IL, Human Kinetics), 37.

TABLE 9.5

Normative Values for the Squat Among Various Athletic Populations

Reference	Population	Gender	Score (mean ± SD, pounds)
BASKETBALL			
Latin et al. 1994	NCAA DI	M	334 ± 81
	G		332 ± 79
	F		356 ± 84
	C		304 ± 70
FOOTBALL			
Garstecki et al. 2004	NCAA DI	M	510 ± 90
	QB		440 ± 99
	RB		513 ± 73
	WR		453 ± 88
	OL		552 ± 75
	TE		510 ± 81
	DL		543 ± 77
	LB		530 ± 81
	DB		458 ± 88
Garstecki et al. 2004	NCAA DII	M	449 ± 90
	QB		394 ± 88
	RB		473 ± 88
	WR		383 ± 77
	OL		488 ± 79
	TE		447 ± 64
	DL		482 ± 79
	LB		460 ± 84
	DB		389 ± 84
SOCCER			
Wisloff et al. 1998	Norwegian elite	M	330 ± 42
VOLLEYBALL			
Fry et al. 1991	NCAA DI	F	180 ± 26

G = guard; F = forward; C = center; QB = quarterback; RB = running back; WR = wide receiver; OL = offensive lineman; TE = tight end; DL = defensive lineman; LB = linebacker; DB = defensive back.

TABLE 9.6

Normative Values for Elite Weightlifters: Front and Back Squat (kg)

	BODY WEIGHT (KG)									
Test	52	56	60	67.5	75	82.5	90	100	110	>110
Back squat	170-190	180-200	200-220	210-230	220-240	230-250	240-260	260-280	270-290	275-300
Front squat	150-165	155-170	170-190	180-200	200-220	210-230	220-240	240-260	245-265	252.5-262.5

Data from Ajan and Baroga, 1988.

TABLE 9.7

Age and Gender Norms for the 1RM Leg Press (Weight Lifted/Body Weight)

WOMEN						
Rating	<20 years	20-29 years	30-39 years	40-49 years	50-59 years	60+ years
Superior	≥1.71	≥1.68	≥1.47	≥1.37	≥1.25	≥1.18
Excellent	1.60-1.70	1.51-1.67	1.34-1.46	1.24-1.36	1.11-1.24	1.05-1.17
Good	1.39-1.59	1.38-1.50	1.22-1.33	1.14-1.23	1.00-1.10	0.94-1.04
Fair	1.23-1.38	1.23-1.37	1.10-1.21	1.03-1.13	0.89-0.99	0.86-0.93
Poor	≤1.22	≤1.22	≤1.09	≤1.02	≤0.88	≤0.85
MEN						
Rating	<20 years	20-29 years	30-39 years	40-49 years	50-59 years	60+ years
Superior	≥2.28	≥2.13	≥1.93	≥1.82	≥1.71	≥1.62
Excellent	2.05-2.27	1.98-2.12	1.78-1.92	1.62-1.81	1.59-1.70	1.50-1.61
Good	1.91-2.04	1.84-1.97	1.66-1.77	1.58-1.68	1.47-1.58	1.39-1.49
Fair	1.71-1.90	1.64-1.83	1.53-1.65	1.45-1.57	1.33-1.46	1.26-1.38
Poor	≤1.70	≤1.63	≤1.52	≤1.44	≤1.32	≤1.25

Adapted, by permission, from The Cooper Institute, 1997, *The Physical Fitness Specialist Certification Manual,* revised 1997 (Dallas, TX: The Cooper Institute for Aerobics Research).

TABLE 9.8

Age and Gender Norms for the 1RM Bench Press (Weight Lifted/Body Weight)

WOMEN						
Rating	<20 years	20-29 years	30-39 years	40-49 years	50-59 years	60+ years
Superior	≥0.78	≥0.81	≥0.71	≥0.63	≥0.56	≥0.55
Excellent	0.66-0.77	0.71-0.80	0.61-0.70	0.55-0.62	0.49-0.55	0.48-0.54
Good	0.59-0.65	0.60-0.70	0.54-0.60	0.51-0.54	0.44-0.48	0.43-0.47
Fair	0.54-0.58	0.52-0.59	0.48-0.53	0.44-0.50	0.40-0.43	0.39-0.42
Poor	≤0.53	≤0.51	≤0.47	≤0.43	≤0.39	≤0.38
MEN						
Rating	<20 years	20-29 years	30-39 years	40-49 years	50-59 years	60+ years
Superior	≥1.34	≥1.32	≥1.12	≥1.00	≥0.90	≥0.82
Excellent	1.20-1.33	1.15-1.31	0.99-1.11	0.89-0.99	0.80-0.89	0.72-0.81
Good	1.07-1.19	1.00-1.14	0.89-0.98	0.81-0.88	0.72-0.79	0.67-0.71
Fair	0.90-1.06	0.89-0.99	0.79-0.88	0.73-0.80	0.64-0.71	0.58-0.66
Poor	≤0.89	≤0.88	≤0.78	≤0.72	≤0.63	≤0.57

Adapted, by permission, from The Cooper Institute, 1997, *The Physical Fitness Specialist Certification Manual,* revised 1997 (Dallas, TX: The Cooper Institute for Aerobics Research).

TABLE 9.9

Normative Values for Relative Bench Press (Weight Lifted/Body Weight) in a General Population

	20-29 YEARS		30-39 YEARS		40-49 YEARS		50-59 YEARS		60+ YEARS	
% rank	M	F	M	F	M	F	M	F	M	F
90	1.48	0.54	1.24	0.49	1.10	0.46	0.97	0.40	0.89	0.41
80	1.32	0.49	1.12	0.45	1.00	0.40	0.90	0.37	0.82	0.38
70	1.22	0.42	1.04	0.42	0.93	0.38	0.84	0.35	0.77	0.36
60	1.14	0.41	0.98	0.41	0.88	0.37	0.79	0.33	0.72	0.32
50	1.06	0.40	0.93	0.38	0.84	0.34	0.75	0.31	0.68	0.30
40	0.99	0.37	0.88	0.37	0.80	0.32	0.71	0.28	0.66	0.29
30	0.93	0.35	0.83	0.34	0.76	0.30	0.68	0.26	0.63	0.28
20	0.88	0.33	0.78	0.32	0.72	0.27	0.63	0.23	0.57	0.26
10	0.80	0.30	0.71	0.27	0.65	0.23	0.57	0.19	0.53	0.25

Adapted, by permission, from V.H. Heyward, 2006, *Advanced fitness assessment & exercise prescription,* 5th ed. (Champaign, IL: Human Kinetics), 123. Data from The Cooper Institute, Dallas, TX, 2005.

TABLE 9.10

Normative Values for the 1RM Bench Press Among Various Athletic Populations

Reference	Population	Gender	Score (mean ± SD, pounds)
BASKETBALL			
Latin et al. 1994	NCAA DI	M	227 ± 42
	G		222 ± 40
	F		229 ± 48
	C		229 ± 37
FOOTBALL			
Garstecki et al. 2004	NCAA DI	M	363 ± 59
	QB		359 ± 48
	RB		385 ± 53
	WR		332 ± 59
	OL		383 ± 62
	TE		378 ± 37
	DL		396 ± 53
	LB		352 ± 53
	DB		312 ± 37
Garstecki et al. 2004	NCAA DII	M	321 ± 57
	QB		284 ± 51
	RB		323 ± 44
	WR		271 ± 44
	OL		352 ± 55
	TE		317 ± 35
	DL		356 ± 46
	LB		321 ± 48
	DB		277 ± 40
RUGBY			
Meir et al. 2001	Australian and English professional	M	
	Forwards		271 ± 26
	Backs		251 ± 37
SOCCER			
Wisloff et al. 1998	Norwegian elite	M	176 ± 33
TEAM HANDBALL			
Hoff & Almasbakk 1995	Norwegian 2nd division	F	121 ± 4
VOLLEYBALL			
Fry et al. 1991	NCAA DI	F	103 ± 18

G = guard; F = forward; C = center; QB = quarterback; RB = running back; WR = wide receiver; OL = offensive lineman; TE = tight end; DL = defensive lineman; LB = linebacker; DB = defensive back.

TABLE 9.11

Percentile Values for the 1RM Bench Press in High School and College Football Players

Percentile	HIGH SCHOOL 14-15 YEARS		HIGH SCHOOL 16-18 YEARS		NCAA DIII		NCAA DI	
	lb	kg	lb	kg	lb	kg	lb	kg
90	243	110	275	125	365	166	370	168
80	210	95	250	114	325	148	345	157
70	195	89	235	107	307	140	325	148
60	185	84	225	102	295	134	315	143
50	170	77	215	98	280	127	300	136
40	165	75	205	93	273	124	285	130
30	155	70	195	89	255	116	270	123
20	145	66	175	80	245	111	255	116
10	125	57	160	73	225	102	240	109
$\bar{X}$	179	81	214	97	287	130	301	137
SD	45	20	44	20	57	26	53	24
n	214		339		591		1189	

TABLE 9.12

Percentile Values for the 1RM Bench Press in NCAA Division I Female Collegiate Athletes

Percentile	BASKETBALL		SOFTBALL		SWIMMING		VOLLEYBALL	
	lb	kg	lb	kg	lb	kg	lb	kg
90	124	56	117	53	116	53	113	51
80	119	54	108	49	109	50	108	49
70	115	52	104	47	106	48	104	47
60	112	51	99	45	101	46	100	45
50	106	48	95	43	97	44	98	45
40	102	46	90	41	94	43	96	44
30	96	44	85	39	93	42	90	41
20	88	40	80	36	88	40	85	39
10	82	37	69	31	78	35	79	36
$\bar{X}$	105	48	94	43	98	45	97	44
SD	18	8	18	8	15	7	14	6
n	120		105		42		67	

Reprinted, by permission, from J. Hoffman, 2006, *Norms for fitness, performance, and health* (Champaign, IL, Human Kinetics), 37.

TABLE 9.13

Modified Unilateral Squat Strength (kg)

Gender	Dominant (SD)	Nondominant (SD)	Difference
Men	107.0 (21.4)	106.0 (21.4)	.9 (4.3)
Women	45.3 (12.5)	45.0 (12.4)	.3 (2.8)

Reprinted from *Journal of Sports Science and Medicine*, Vol 4, K. McCurdy and G. Langford, "Comparison of unilateral squat strength between the dominant and non-dominant leg in men and women," pgs. 153-159, copyright 2005, with permission from the J Sports Sci & Med.

TABLE 9.14

Standing Long Jump Test Values

Reference	Population	Test variable	Score (mean ± SD)
Robertson & Fleming 1987	Females (n = 4) Males (n = 2)	Average distance	215.2 ± 25.3 cm
		Peak vertical force	2.05 ± 0.13 × body weight
		Peak horizontal force	0.65 ± 0.07 × body weight
Markovic et al. 2004	Physically active college-age men (n = 93)	Average distance	251.5 ± 15.0 cm
Koch et al. 2003	Athletic men (n = 8) (179 ± 3 cm average height)	Average distance	2.43 ± 0.14 m
	Athletic women (n = 3) (166 ± 7 cm average height)	Average distance	2.28 ± 0.16 m
	Nonathletic men (n = 8) (179 ± 7 cm average height)	Average distance	2.33 ± 0.16 m
	Nonathletic women (n = 13) (166 ± 8 cm average height)	Average distance	1.61 ± 0.28 m

TABLE 9.15

Percentile Ranks for Standing Long Jump in Elite Male and Female Athletes

	MALES		FEMALES	
% rank	in.	cm	in.	cm
90	148	375	124	315
80	133	339	115	293
70	122	309	110	279
60	116	294	104	264
50	110	279	98	249
40	104	264	92	234
30	98	249	86	219
20	92	234	80	204
10	86	219	74	189

Reprinted, by permission, from J. Hoffman, 2006, *Norms for fitness, performance, and health* (Champaign, IL: Human Kinetics), 58.

TABLE 9.16

Rankings for Standing Long Jump in 15- and 16-Year-Old Male and Female Athletes

	MALES		FEMALES	
Category	in.	cm	in.	cm
Excellent	79	201	65	166
Above average	73	186	61	156
Average	69	176	57	146
Below average	65	165	53	135
Poor	<65	<165	<53	<135

Reprinted, by permission, from J. Hoffman, 2006, *Norms for fitness, performance, and health* (Champaign, IL: Human Kinetics), 58.

TABLE 9.17

Studies, Scores, and Reliability of the Single-Leg Hop for Distance Test

Reference	Subjects	Test variable	Score (mean ± SD) Distance in cm (SD)	Reliability (ICC)
Ageberg et al. 1998	75 normal subjects			0.96
Bandy et al. 1994	18 normal subjects	Dominant leg		0.93

Barber et al. 1990	93 normal subjects	High-level females
		   Dominant leg
		   Nondominant leg
			
			121.5 ± 18.5
			122.0 ± 21.0	

		Medium-level females
		   Dominant leg
		   Nondominant leg
			
			117.8 ± 21.3
			113.7 ± 19.3	

		High-level males
		   Dominant leg
		   Nondominant leg
			
			187.8 ± 25.5
			189.6 ± 27.6	

		Medium-level males
		   Dominant leg
		   Nondominant leg
			
			149.6 ± 17.3
			150.7 ± 16.2	

		Male soccer
		   Dominant leg
		   Nondominant leg
			
			204.0 ± 14.9
			202.3 ± 8.8	

Bolgla et al. 1997	20 subjects: 5 males, 15 females	Dominant leg		0.96
Booher et al. 1993	18 subjects: 4 males, 14 females	Dominant leg		0.99
Brosky et al. 1999	15 males with unilateral ACL reconstruction	Uninvolved Involved	173.5 ± 21.1 173.1 ± 18.5	0.88 to 0.97 Specifics not reported

DeCarlo & Sell 1997	889 healthy subjects: 578 males, 311 females	Men
		   Right
		   Left
		% difference
			
			153.9 ± 27.5
			155.1 ± 28.5
			99.7 ± 8.32	

		Women
		   Right
		   Left
		% difference
			
			122.5± 19.5
			119.9 ± 20.5
			102.8 ± 9.25	

Greenberger & Paterno 1994	20 normal subjects	Dominant leg Nondominant leg		0.92 0.96
Greenberger & Paterno 1995	20 normal subjects	Dominant leg Nondominant leg	147.4 ± 26.11 143.4 ± 27.00	

Reference	Subjects	Test variable	Score (mean ± SD) Distance in cm (SD)	Reliability (ICC)

Gauffin et al. 1990	15 patients with chronic ACL rupture, braced and unbraced, control	ACL deficient
		   Brace
		   No brace
		Uninjured knee
		Control group
			
			158 ± 16
			157 ± 16
			164 ± 13
			177 ± 13	

Hu et al. 1992	30 normal subjects	Bilateral legs		0.79 to 0.96 Specifics not reported
Itoh et al. 1998	50 patients with unilateral ACL deficiency: 23 males, 27 females	Males, dominant leg Females, dominant leg Males, nondominant leg Females, nondominant leg	193 ± 0.19 149 ± 0.14 184 ± 0.18 142 ± 0.14	

Keays et al. 2001	31 patients with unilateral ACL reconstruction with hamstring autograft	Presurgery
		   Injured leg
		   Uninjured leg
		Postsurgery (6 months)
		   Injured leg
		   Uninjured leg
			
			122.87 ± 37.53
			149.63 ± 26.91
			
			136.35 ± 28.87
			155.09 ± 23.49	

Kramer et al. 1992	38 patients with unilateral ACL: 22 males, 16 females		126.5	Men, Inv = 0.85 Men, Un = 0.77 Women, Inv = 0.94 Women, Un = 0.90
Manske et al. 2003	28 normal subjects	Dominant leg Nondominant leg	126.6 ± 21.6 127.8 ± 28.2	0.96 0.97
Negrete & Brophy 2003	60 normal subjects			0.85
Ross et al. 2002	18 normal subjects			0.92
Vandermeulen et al. 2000	46 normal subjects			0.84 to 0.92

ACL = anterior cruciate ligament.

TABLE 9.18

Reliability of the Modified Single-Leg Hop Onto a Force Plate

Reference	Subjects	Test parameters	Reliability (ICC) and stabilization times (SD)
Colby et al. 1999	25 normal subjects 11 ACL reconstructed 13 ACL deficient	Anterior-posterior force	
		Dominant leg	0.958 ± 0.988
		Nondominant leg	0.925 ± 0.988
		Medial-lateral force	
		Dominant leg	0.872 ± 0.951
		Nondominant leg	0.875 ± 0.966
		Vertical force	
		Dominant leg	0.899 ± 0.983
		Nondominant leg	0.971 ± 0.984
		Anterior-posterior center of pressure	
		Dominant leg	0.682 ± 0.062
		Nondominant leg	0.752 ± 0.754
		Medial-lateral center of pressure	
		Dominant leg	0.53 ± 0.366
		Nondominant leg	0.715 ± 0.202

ACL = anterior cruciate ligament; ICC = Intraclass correlation coefficient; SD = Standard deviation

TABLE 9.19

Correlations of the Single-Leg Hop to Various Parameters

Authors	Subjects	Conditions	Correlations
Barber et al. 1990	93 normal subjects		
Delitto et al. 1993	30 patients with unilateral ACL reconstruction	Bilateral concentric and eccentric quadriceps peak torque and work at 60°/s and 120°/s	Concentric quad peak torque and work ($r = 0.038$ to 0.46) Eccentric quad peak torque and work ($r = 0.09$ to 0.27)
Greenberger & Paterno 1995	20 normal subjects	Concentric quadriceps peak torque at 240°/s	Dominant peak torque ($r = 0.782$) Nondominant peak torque ($r = 0.649$)
Sernert et al. 1999	527 patients with unilateral ACL reconstruction	Lysholm Activity Score	Lysholm Activity Score ($p = 0.36$)
Swarup et al. 1992	30 normal subjects	Bilateral concentric and eccentric quadriceps peak torque and work at 60°/s and 120°/s	Males, left leg concentric quad peak torque at 60° and 120° and work at 120° ($r = 0.72$ to 0.89) Males, right leg concentric quad peak torque at 60° and 120° and work at 120° ($r = 0.56$ to 0.73) Females, concentric quad peak torque at 60° and 120° and work at 120° ($r = 0.57$ to 0.63)
Tegner et al. 1986		Bilateral concentric quadriceps peak torque at 120°/s	
Wiklander & Lysholm 1987		Bilateral concentric peak torque at 180°/s	
Wilk et al. 1994	50 ACL reconstruction patients	Quadriceps peak torque at 180°/s, 300°/s, and 450°/s	Concentric quad peak torque ($r = 0.41$ to 0.62) Concentric quad extension acceleration velocity ($r = 0.06$ to 0.41) Concentric quad deceleration velocity ($r = 0.09$ to 0.17)

ACL = anterior cruciate ligament.

TABLE 9.20

Percentile Ranks for Vertical Jump Heights in Youth

% rank	7-8 YEARS		9-10 YEARS		11-12 YEARS	
	in.	cm	in.	cm	in.	cm
90	9.6	24.4	11.5	29.2	16.5	41.9
80	9.3	23.6	11.0	27.9	14.3	36.3
70	8.7	22.1	10.4	26.4	12.3	31.2
60	8.1	20.6	9.9	25.1	11.8	30.0
50	8.0	20.3	9.5	24.1	10.5	26.7
40	7.7	19.6	9.0	22.9	10.0	25.4
30	7.5	19.1	8.6	21.8	9.6	24.4
20	7.1	18.0	7.8	19.8	8.8	22.4
10	6.9	17.5	7.0	17.8	6.8	17.3
$\bar{X}$	8.1	20.6	9.3	23.6	11.2	28.4
SD	1.0	2.5	1.7	4.3	3.5	8.9
n	26		67		74	

% rank	MALES 13-14 YEARS		FEMALES 13-14 YEARS		MALES 15-16 YEARS		FEMALES 14-15 YEARS		MALES 17-18 YEARS	
	in.	cm	in.	cm	in.	cm	in.	cm	in.	cm
90	21.0	53.3	17.0	43.2	27.0	68.6	18.5	47.0	28.2	71.6
80	20.0	50.8	16.0	40.6	24.0	61.0	17.5	44.5	26.0	66.0
70	19.0	48.3	16.0	40.6	22.5	57.2	16.9	42.9	25.0	63.5
60	18.4	46.7	15.0	38.1	22.0	55.9	16.0	40.6	23.8	60.5
50	17.0	43.2	14.5	36.8	20.5	52.1	15.5	39.4	22.0	55.9
40	16.0	40.6	14.0	35.6	20.0	50.8	14.9	37.8	20.2	51.3
30	15.0	38.1	14.0	35.6	20.0	20.8	14.1	35.8	19.4	49.3
20	13.8	35.1	13.5	34.3	17.0	43.2	13.2	33.5	18.6	47.2
10	12.3	31.2	13.0	33.0	17.0	43.2	10.0	25.4	18.0	45.7
$\bar{X}$	16.8	42.7	14.6	37.1	20.9	53.1	15.2	38.6	22.6	57.4
SD	3.4	8.6	1.5	3.8	3.4	8.6	2.7	6.9	3.8	9.7
n	42		19		29		16		27	

Reprinted, by permission, from J. Hoffman, 2006, *Norms for fitness, performance, and health* (Champaign, IL, Human Kinetics), 59.

TABLE 9.21

Percentile Ranks for Vertical Jump (No Step) in Basketball Players

% rank	HS 14 YEARS in.	HS 14 YEARS cm	HS 15 YEARS in.	HS 15 YEARS cm	HS 16 YEARS in.	HS 16 YEARS cm	HS 17 YEARS in.	HS 17 YEARS cm
90	25.6	65.0	27.1	68.8	29.0	73.7	28.3	71.9
80	23.4	59.4	25.0	63.5	27.5	69.9	26.5	67.3
70	22.5	57.2	24.0	61.0	25.7	65.3	24.5	62.2
60	21.6	54.9	23.0	58.4	24.7	62.7	24.0	61.0
50	21.0	53.3	23.0	58.4	24.0	61.0	24.0	61.0
40	20.9	53.1	22.0	55.9	23.0	58.4	23.5	59.7
30	20.3	51.6	21.5	54.6	22.4	56.9	22.9	58.2
20	18.0	45.7	20.5	52.1	20.9	53.1	21.6	54.9
10	15.4	39.1	20.0	50.8	19.5	49.5	21.0	53.3
$\bar{X}$	21.0	53.3	23.1	58.7	24.0	61.0	24.0	61.0
SD	3.1	7.9	3.0	7.6	3.9	9.9	2.3	5.8
n	21		87		58		22	

% rank	NCAA DI MALES in.	NCAA DI MALES cm	NCAA DI FEMALES in.	NCAA DI FEMALES cm	NBA in.	NBA cm
90	30.5	77.5	21.6	54.9	31.2	79.2
80	30.0	76.2	20.1	51.1	29.5	74.9
70	28.5	72.4	19.7	50.0	28.4	72.1
60	28.0	71.1	18.5	47.0	27.5	69.9
50	27.5	69.9	18.0	45.7	27.0	68.6
40	26.8	68.1	17.5	44.5	26.2	66.5
30	26.0	66.0	16.5	41.9	24.6	62.5
20	25.5	64.8	15.9	40.4	23.6	59.9
10	24.5	62.2	14.5	36.8	22.4	56.9
$\bar{X}$	27.7	70.4	18.0	45.7	26.7	67.8
SD	2.4	6.1	2.5	6.4	3.3	8.4
n	138		118		40	

HS = high school.

Reprinted, by permission, from J. Hoffman, 2006, *Norms for fitness, performance, and health* (Champaign, IL, Human Kinetics), 61.

TABLE 9.22

Percentile Ranks for Vertical Jump in Volleyball Players

AGE	SEX	SCORE										
		100	90	80	70	60	50	40	30	20	10	0
9-11	Boys and girls	16	15	14	12	11	10	9	7	4	2	0
12-14	Boys	20	18	17	16	14	13	11	9	5	2	0
	Girls	16	15	14	13	12	11	10	8	4	2	0
15-17	Boys	25	24	23	21	19	16	12	8	5	2	0
	Girls	17	16	15	14	13	11	8	6	3	2	0
18-34	Men	26	25	24	23	19	16	13	9	8	2	0
	Women	14	13	13	12	10	8	6	4	2	1	0

Reprinted, by permission, from H.T. Friermood, 2004, Volleyball skills contest for olympic development. In *United States Volleyball Association, annual official volleyball rules and reference guide of the U.S. Volleyball Association.* Courtesy of USA Volleyball.

TABLE 9.23

Percentile Ranks for Vertical Jump (No Step) in NCAA Division I Female Athletes

% rank	VOLLEYBALL		SOFTBALL		SWIMMING	
	in.	cm	in.	cm	in.	cm
90	20.0	50.8	18.5	47.0	19.9	50.5
80	18.9	48.0	17.0	43.2	18.0	45.7
70	18.0	45.7	16.0	40.6	17.4	44.2
60	17.2	44.5	15.0	38.1	16.1	40.9
50	17.0	43.2	14.5	36.8	15.0	38.1
40	16.7	42.4	14.0	35.6	14.5	36.8
30	16.5	41.9	13.0	33.0	13.0	33.0
20	16.0	40.6	12.0	30.5	12.5	31.8
10	15.5	39.4	11.0	27.9	11.6	29.5
$\overline{X}$	17.3	43.9	14.6	37.1	15.3	38.9
SD	2.1	5.3	2.9	7.4	3.0	7.6
n	90		118		40	

Reprinted, by permission, from J. Hoffman, 2006, *Norms for fitness, performance, and health* (Champaign, IL, Human Kinetics), 62.

TABLE 9.24

Canadian Vertical Jump
Normative Data by Gender and Age (Mean ± SEM)

Age group (year)	n	Vertical jump (cm)	Vertical jump (in.)
FEMALES			
15-19	59	34 ± 1	13.4 ± .39
20-29	83	30 ± 1	11.8 ± .39
30-39	56	29 ± 1	11.4 ± .39
40-49	47	23 ± 1	9.1 ± .39
50-59	47	18 ± 1	7.1 ± .39
60-69	20	11 ± 1	4.3 ± .39
MALES			
15-19	54	48 ± 1	18.9 ± .39
20-29	73	50 ± 1	19.7 ± .39
30-39	44	43 ± 1	16.9 ± .39
40-49	27	35 ± 1	13.8 ± .39
50-59	36	28 ± 1	11 ± .39
60-69	25	24 ± 1	9.5 ± .39

Adapted, by permission, from N. Payne et al., 2000, "Canadian musculoskeletal fitness norms," *Canadian Journal of Applied Physiology* 25:430-442.

TABLE 9.25

Normative Data for Vertical Jump (cm) for Male and Female Canadians

Percentile range	AGE					
	15-19	20-29	30-39	40-49	50-59	60-69
FEMALES						
91-100	≥42	≥41	≥38	≥33	≥27	≥21
81-90	40-41	38-40	36-37	31-32	25-26	19-20
71-80	38-39	36-37	34-35	29-30	23-24	17-18
61-70	36-37	34-35	32-33	27-28	21-22	15-16
51-60	34-35	31-33	30-31	25-26	19-20	13-14
41-50	32-33	29-30	28-29	23-24	16-18	11-12
31-40	30-31	27-28	26-27	21-22	13-15	9-10
21-30	28-29	25-26	24-25	18-20	10-12	7-8
11-20	26-27	20-24	20-23	15-17	6-9	4-6
0-10	≤25	≤19	≤19	≤14	≤5	≤3

(continued)

TABLE 9.25 *(continued from previous page)*

Percentile range	AGE					
	15-19	20-29	30-39	40-49	50-59	60-69
MALES						
91-100	≥58	≥61	≥55	≥52	≥48	≥35
81-90	56-57	58-60	52-54	43-51	41-47	33-34
71-80	54-55	56-57	49-51	39-42	37-40	31-32
61-70	51-53	54-55	46-48	36-38	34-36	29-30
51-60	48-50	51-53	43-45	34-35	31-33	27-28
41-50	46-47	48-50	40-42	32-33	28-30	25-26
31-40	44-45	45-47	37-39	30-31	25-27	23-24
21-30	42-43	42-44	31-36	26-29	18-24	18-22
11-20	39-41	39-41	24-30	22-25	11-17	13-17
0-10	≤38	≤38	≤23	≤21	≤10	≤12

Adapted, by permission, from N. Payne et al., 2000, "Canadian musculoskeletal fitness norms," *Canadian Journal of Applied Physiology* 25:430-442.

TABLE 9.26

Vertical Jump Deciles for Female and Male College-Age Subjects (*n* = 152)

Deciles	Females Vertical jump (cm)	Males Vertical jump (cm)
90	53.34	74.93
80	48.76	71.12
70	45.72	66.04
60	41.91	62.23
50	40.64	58.42
40	36.83	54.86
30	33.02	53.34
20	30.48	49.53
10	27.94	43.18

Adapted, by permission, from K. Pauole et al., 2000, "Reliability and validity of the T-test as a measure of agility, leg power, and leg speed in college-aged men and women," *Journal of Strength and Conditioning Research* 14:443-450.

TABLE 9.27

Vertical Jump Normative Values for Athletes in Various Sports

Study	Gender	Population	*n*	Vertical jump (cm)
BASKETBALL				
Hoffman et al. 1991	M	NCAA Division I	9	64.3 ± 7.9
Hoffman et al. 1996	M	NCAA Division I	29	67.3 ± 6.0
Latin et al. 1994	M	NCAA Division I	437	Centers: 66.8 ± 10.7 Forwards: 71.4 ± 10.4 Guards: 73.4 ± 9.6
LaMonte et al. 1999	F	NCAA Division I	9 19 18	Centers: 43.5 ± 4.5 Forwards: 49.4 ± 11.1 Guards: 49.4 ± 6.2
Soares et al. 1986	M	Brazilian National	21	Centers: 55.9 ± 8.1 Forwards: 66.8 ± 8.3 Guards: 61.6 ± 8.5
Smith & Thomas 1991	F	Canadian National		Centers: 42.0 ± 3.0 Forwards: 44.5 ± 4.4 Guards: 48.9 ± 4.9
Woolstenhulme et al. 2004	F	NCAA Division I	18	1-leg takeoff jump: 33.3 ± 3.8 2-leg takeoff jump: 49.5 ± 4.8
VOLLEYBALL				
Fry et al. 1991	F	NCAA Division I		48.0 ± 4.2
Stockbrugger & Haennel 2003	M	Competitive players	20	62.0 ± 7
FOOTBALL				
Garstecki et al. 2004	M	NCAA Division I	26 50 48 124 26 101 75 99	QB: 80.7 ± 6.4 RB: 85.9 ± 7.7 WR: 87.4 ± 7.0 OL: 68.8 ± 6.2 TE: 79.6 ± 7.2 DL: 77.9 ± 8.2 LB: 83.2 ± 7.8 DB: 87.8 ± 7.8
Garstecki et al. 2004	M	NCAA Division II	18 35 34 88 17 70 53 70	QB: 70.3 ± 9.3 RB: 74.2 ± 11.0 WR: 77.8 ± 12.1 OL: 60.4 ± 8.6 TE: 70.1 ± 8.7 DL: 66.9 ± 11.3 LB: 72.4 ± 10.8 DB: 78.0 ± 10.3

(continued)

TABLE 9.27 *(continued from previous page)*

Study	Gender	Population	*n*	Vertical jump (cm)
WRESTLING				
Callan et al. 2000	M	National	8	60 ± 10
Kraemer & Gotshalk 2000	M	NCAA Division I		23.6 ± 3.9 in.
Stockbrugger & Haennel 2003	M	Competitive players	20	52 ± 8
WEIGHTLIFTING				
Fry et al. 2003	M	National	6	60.8 ± 3.9
SOCCER				
Kirkendall 2000	F	American elite youth and adults*		U12: 40.3 U13: 40.8 U14: 42.8 U15: 44.9 U16: 45.6 U17: 49.6 U18: 43.1
Kirkendall 2000	M	National		U13: 47.2 U14: 53.8 U15: 61.0 U16: 65.2

* = one step was permitted. QB = quarterback; RB = running back; WR = wide receiver; OL = offensive lineman; TE = tight end; DL = defensive lineman; LB = linebacker; DB = defensive back.

TABLE 9.28

Reliability of the 6 m Timed Hop

Authors	Subjects	Conditions	TRT reliability	Validity	Correlation
Bandy et al. 1994	18 subjects, gender NR	2 trials; 10 ft rather than 6 m	0.92	NR	NR
Bolgla & Keskula 1997	20 subjects: 5 men, 15 women	3 trials	0.66	NR	NR
Booher et al. 1993	18 subjects: 4 men, 14 women	1 trial	0.77	NR	NR
Manske et al. 2003	28 subjects, gender NR	3 trials	0.92 DL 0.96 NDL	NR	NR
Reid et al. 2007	42 subjects who had undergone ACL reconstruction	3 trials	0.82	NR	0.42 with GROC 0.28 with LEFS
Ross et al. 2002	18 subjects, all men	3 trials	0.92	NR	NR

TRT = test-retest; GROC = global rating of change self-report measure; LEFS = Lower Extremity Functional Scale; NR = not reported; DL = dominant leg; NDL = nondominant leg

TABLE 9.29

Studies, Scores, and Reliability of the Single-Leg Crossover Hop for Distance

Reference	Subjects	Test variable	Score (mean ± SD) Distance in cm (SD)	Reliability (ICC)
Clark et al. 2002	12 subjects: 4 male, 8 female	Maximal hop distance (4 hops)	601.1 (117.6)	0.94
Gaunt & Curd 2001	201 high school athletes	Maximal hop distance (3 hops)	Girls (n = 85), 381.7 (67.8) Boys (n = 116), 505.5 (101.4)	
		Limb symmetry index	Girls, 91.0 (0.7) Boys, 93.0 (0.6)	
Goh & Boyle 1997	10 subjects	Distance hopped		0.85
Wilk et al. 1994	50 ACL reconstruction patients: 29 males, 21 females	Maximal hop distance (3 hops)	Noninvolved, 187.6 (36.1) Involved, 161.9 (39.7)	
Reid et al. 2007	42 ACL reconstruction patients: 23 males, 19 females	Maximal hop distance (3 hops)	Noninvolved, 376.1 (83.2) Involved, 336.9 (87.9) at 22 weeks post for females	0.84
			Noninvolved, 431.0 (89.4) Involved, 377.2 (88.3) at 22 weeks post for males	
Ross et al. 2002	18 subjects, all males	Maximal hop distance		0.93

ACL = anterior cruciate ligament.

TABLE 9.30

Correlations of the Single-Leg Crossover Hop for Distance With Isokinetic Tests

Speed	Knee extension peak	Knee extension acceleration	Knee extension deceleration
180	$P = 0.0001; r = 0.69$	$P = 0.05; r = 0.53$	$P = 0.0001; r = 0.51$
300	$P = 0.001; r = 0.64$	$P = 0.05; r = 0.54$	$P = 0.05; r = 0.49$
450	$P = 0.001; r = 0.53$	$P = 0.30; r = 0.16$	$P = 0.17; r = 0.23$

Data from Wilk et al., 1994.

TABLE 9.31

Hop-and-Stop Test Results: Normal Clients ($n = 100$) and Patients ($n = 20$)

	Hop % height*	Hop symmetry	Stop-to-hop ratio*	Stop symmetry
Normal women ($n = 50$)	70.48 ± 10.88 (65%)	93.98 ± 4.42	113.67 ± 9.54	93.57 ± 5.37
Normal men ($n = 50$)	83.85 ± 7.57	94.26 ± 5.28	110.04 ± 8.15	93.11 ± 5.95
Combined ($n = 100$)	N/A	94.12 ± 4.84 (90%)	111.86 ± 8.85 (105%)	93.34 ± 5.66 (90%)
Symptomatic women ($n = 9$)	54.04 ± 6.06#	82.72 ± 1.59	99.83 ± 15.19	77.91 ± 3.74
Symptomatic men ($n = 11$)	66.82 ± 14.59#	80.29 ± 11.77	91.99 ± 9.73	75.92 ± 13.15
Combined ($n = 20$)		81.00 ± 9.86#	94.29 ± 11.67#	76.50 ± 11.10#

Values shown are mean ± standard deviation; numbers in parentheses are the rounded criterion performance scores.

*Hop % height and stop-to-hop ratio derived from mean scores of both legs in normals.

#Significantly different from normals.

Adapted, by permission, from P.M. Juris et al., 1997, "A dynamic test of lower extremity function following anterior cruciate ligament reconstruction and rehabilitation," *Journal of Orthopedic Sports Physical Therapy* 26:184-191.

TABLE 9.32

Predictive Value of Hop-and-Stop Test

	Hop % height	Hop symmetry	Stop-to-hop ratio	Stop symmetry
Normal (specificity)	78	92	81	89
Symptomatic (sensitivity)	90	95	85	95

Specificity = percentage of normal subjects who pass criteria; sensitivity = percentage of dysfunctional subjects who fail to meet criteria.

Reprinted, by permission, from P.M. Juris et al., 1997, "A dynamic test of lower extremity function following anterior cruciate ligament reconstruction and rehabilitation," *Journal of Orthopedic Sports Physical Therapy* 26:184-191.

TABLE 9.33

Normative Values for Selected Tests

Test	Normal values Male (*) (*n* = 23)	Normal values Female (*) (*n* = 37)	Intertester (#) ICC (SEM) Athlete	Nonathlete	Intratester (#) ICC (SEM) Athlete	Nonathlete
Figure 8 (s) Dominant leg Nondominant leg	11.36 ± 1.30 11.39 ± 1.38	12.47 ± 0.89 12.64 ± 0.88	(*n* = 23) 0.99 (0.32)	(*n* = 23) 0.99 (0.23)	(*n* = 23) 0.85 (0.91)	(*n* = 23) 0.92 (0.82)
Up-down (s) Dominant leg Nondominant leg	7.60 ± 1.13 7.62 ± 1.06	8.68 ± 0.78 8.65 ± 0.75	(*n* = 22) 0.96 (0.42)	(*n* = 15) 0.99 (0.36)	(*n* = 22) 0.84 (0.81)	(*n* = 15) 0.92 (0.96)
Side hop (s) Dominant leg Nondominant leg	7.36 ± 1.51 7.40 ± 1.56	8.20 ± 0.80 8.21 ± 0.82	(*n* = 24) 0.97 (0.41)	(*n* = 20) 0.99 (0.27)	(*n* = 24) 0.48 (1.43)	(*n* = 20) 0.88 (0.95)
Single hop (m) Dominant leg Nondominant leg	1.93 ± 0.19 1.84 ± 0.18	1.49 ± 0.14 1.42 ± 0.14	0.97 (0.41)	0.99 (0.27)	0.48 (1.43)	0.88 (0.95)

* Itoh et al. 1998; # Ortiz et al. 2005; significant differences (*P* < .05) existed for all tests for both males and females.

TABLE 9.34

Values for Left–Right Difference in the FAT: Comparison Between Normal and ACL-Deficient Clients

Test	Normal (*n* = 60)	ACL deficient (*n* = 50)	Sensitivity (%)	Specificity (%)
Figure 8 (s)	0.35 ± 0.23	1.45 ± 1.12	68	98
Up-down (s)	0.30 ± 0.21	1.67 ± 2.41	58	97
Side hop (s)	0.32 ± 0.23	0.87 ± 0.86	44	95
Single hop (m)	0.08 ± 0.06	0.18 ± 0.15	42	98

Sensitivity: percent probability that the tests would demonstrate abnormal lower limb symmetry in ACL-deficient patients; specificity: percent probability that the tests would demonstrate normal lower limb symmetry in normal subjects.

ACL = anterior cruciate ligament; FAT = functional ability tests.

Itoh et al. 1998; Ortiz et al. 2005.

TABLE 9.35

Hexagon Test (Bilateral Leg Jump): Normative Values for Athletes

Sport	Time (s)
Competitive college athletes (men)	12.3
Competitive college athletes (women)	12.9
Recreational college athletes (men)	12.3
Recreational college athletes (women)	13.2

Values are either means or 50th percentiles.

Reprinted, by permission, from NSCA, 2008, Administration, scoring, and interpretation of selected tests, by E. Harman, J. Garhammer, and C. Pandorf. In *Essentials of strength training and conditioning,* 3 rd ed., edited by T.R. Baechle and R.W. Earle (Champaign, IL: Human Kinetics), 278.

TABLE 9.36

Mean Scores (±SD) for the Hexagon Test (Bilateral Leg Jumps) in Female and Male College-Age Clients

Group	n	T-test norms (s)
FEMALE SUBJECTS		
Recreational sport	52	13.21 ± 1.68
College athletes	56	12.87 ± 1.48
MALE SUBJECTS		
Recreational sport	58	12.33 ± 1.47
College athletes	47	12.29 ± 1.39

Reprinted, by permission, from K. Pauole et al., 2000, "Reliability and validity of the T-test as a measure of agility, leg power, and leg speed in college-aged men and women," *Journal of Strength Conditioning Research* 14:443-450.

TABLE 9.37

Hexagon (Bilateral Leg Jump) Deciles for Female and Male College-Age Subjects ($n = 152$)

Deciles	Females hexagon (s)	Males hexagon (s)
90	11.17	10.68
80	12.01	11.32
70	12.53	11.87
60	13.02	12.14
50	13.22	12.58
40	13.71	13.20
30	14.23	13.84
20	14.94	14.25
10	16.34	14.96

Reprinted, by permission, from K. Pauole et al., 2000, "Reliability and validity of the T-test as a measure of agility, leg power, and leg speed in college-aged men and women," *Journal of Strength Conditioning Research* 14:443-450.

chapter 10

Speed, Agility, and Quickness Testing

The tests described in this chapter assess the client's ability to move with precision and speed. Speed is the ability to achieve high velocity (Baechle & Earle 2000). Agility is the ability to explosively stop, change direction, and accelerate again. These components are very important in many sports, such as football, basketball, and soccer. The ability of a wide receiver to get around the defensive back is dependent on these motor skills and abilities. Speed, agility, and quickness are very often interrelated, especially in sporting activities. These tests are done in an effort to assess clients' ability to react and move quickly through a test sequence. Tests such as the Zigzag Run Test and the Lower Extremity Functional Test (LEFT) are designed to include components of speed, agility, and quickness in multiple planes of movement. These tests are typically longer and more difficult than most of the traditional testing for these categories.

T-TEST

▶ **Purpose:** To test multidirectional speed, agility, and body control.

▶ **Equipment:** Flat nonslip surface, four cones, stopwatch, measuring tape.

Procedure (Seminick 1990; Fry et al. 1991)

1. Instruct the client to sprint 10 yd (9.1 m) forward, touching the right hand to the base of the center cone (cone B).

2. Have the client then side shuffle to the left 5 yd (4.6 m); the left hand touches the base of cone C.

3. Have the client shuffle right 10 yd; the right hand touches the base of cone D.

4. Have the client again side shuffle left to the center cone; the left hand touches the base of cone B.

5. Have client finally backpedal past the starting cone, and stop the clock.

6. Perform two trials and record the best time to the nearest tenth of a second.

7. Allow the client to rest for 1 to 2 min between the two trials.

8. Use the following disqualification criteria: Client doesn't touch the base of a cone, crosses feet when shuffling, or fails to face the front at all times.

Analysis and Interpretation of Data

Normative values and percentile ranks for various populations for this test are listed in tables 10.1 through 10.6.

Statistics

Intraclass correlation coefficient (ICC) values ranged from 0.94 to 0.98 (Pauole et al. 2000).

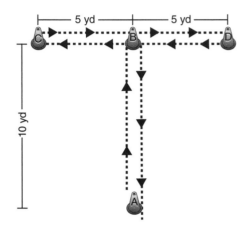

PRO AGILITY (5-10-5) TEST

▶ **Purpose:** To test multidirectional speed, agility, and body control.

▶ **Equipment:** Flat nonslip surface, stopwatch, measuring tape.

Procedure (Harman et al. 2000)

1. Have the client start in a two-point stance straddling the starting line.
2. Instruct the client on the "Go" command to turn and sprint to the right, touching the line 5 yd (4.6 m) away with the right hand.
3. Have the client then turn to the left, sprint 10 yd (9.1 m), and touch the far line with the left hand.
4. Have the client then turn back to the right and sprint 5 yd through the start/finish line, and stop the clock.
5. Use the disqualification criterion of not touching line with alternate hand.

Analysis and Interpretation of Data

Percentile ranks and normative values for various athletic populations are listed in tables 10.7 through 10.9.

Statistics

There are no known reliability or validity data for this assessment.

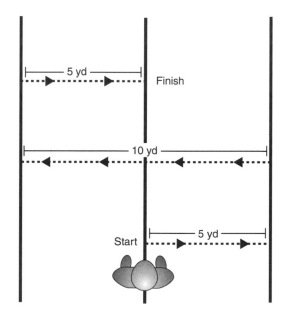

505 AGILITY TEST

► **Purpose:** To test multidirectional speed, agility, and body control.

► **Equipment:** Flat nonslip surface with a marked starting line, a turning point line, and a light gate or two pieces of colored tape for a start/stop gate; standard stopwatch.

Procedure

1. Instruct the client to begin at the starting line and sprint through the light gate to the zero line.
2. At this position, require the client to turn on either the left or the right foot and then accelerate off the line back through the light gate.
3. Tell the client not to slow down until after he or she has passed through the light gate for the second time.
4. Have the client complete three trials, turning on the preferred foot.

5. Alternatively, you may ask the client to perform three trials turning on the left foot and three trials turning on the right foot, for a total of six trials.
6. Implement appropriate rest between repetitions.
7. Record to the nearest tenth of a second the time taken to cover the 10 m distance from the light gate to the turning point and back.

Analysis and Interpretation of Data

■ The fastest time is recorded as the best score.
■ Normative values for this test are listed in table 10.10.

Statistics

There are no known reliability or validity data for this assessment.

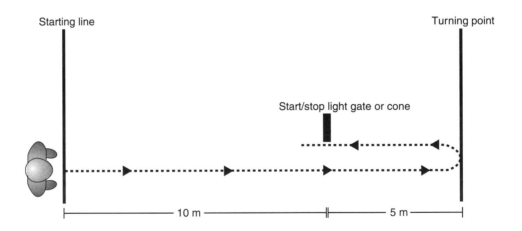

EDGREN SIDE-STEP TEST

▶ **Purpose:** To test lateral speed, agility, and body control.

▶ **Equipment:** Flat nonslip surface or other flooring, 12 ft (3.7 m) wide, marked every 3 ft (0.9 m) with colored tape; standard stopwatch.

Procedure (Semenick 2000)

1. Have the client stand in a two-point stance, straddling the center line.
2. Instruct the client to sidestep to the right until his or her right foot has touched or crossed the right outside line.
3. Instruct the client to sidestep to the left until his or her left foot has touched or crossed the left outside line.
4. Have the client continue to sidestep back and forth to the outside lines as rapidly as possible for 10 s.
5. Ensure that the client faces forward and side shuffles for the entire test.

Analysis and Interpretation of Data

■ Each completion of a 3 ft (0.9 m) increment (from center line to first increment, from first increment to outside line, from outside line back to first increment, etc.) counts as 1 point.

■ Disqualification criteria include a 1-point penalty for each time one foot crosses the other and for each failure to get the proper foot on or across the outside marker.

Statistics

There are no known reliability or validity data for this assessment.

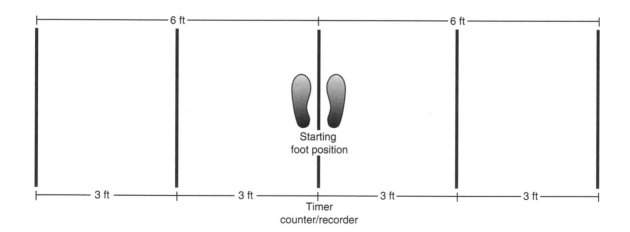

THREE-CONE DRILL TEST

▶ **Purpose:** To test multidirectional speed, agility, and body control.

▶ **Equipment:** Three cones, nonslip testing surface, standard stopwatch or electronic testing device.

Procedure

1. Set up three cones in an upside-down "L" configuration as in diagram, placing each cone 5 yd (4.6 m) apart from the others.

2. Instruct the client to sprint from the starting line at cone A to cone B, touching the cone.

3. Have the client reverse direction and sprint back to cone A, then change direction and sprint around cone B to cone C.

4. Have the client then circle cone C to his or her left and sprint back around cone B and past the original starting line at cone A.

5. You can allow the client to run through the routine as a warm-up and to ensure complete understanding of procedures.

6. Record the time to the nearest 0.01 s.

Analysis and Interpretation of Data

◾ National Football League (NFL) drafted rookies (males) (McGee & Burkett 2003):

- Rounds 1 + 2: 7.23 ±@ 0.41.
- Rounds 6 + 7: 7.46 ±@ 0.46.
- Normative values and percentile ranks for NFL combine times of college football players are listed in table 10.11.

Statistics

There are no known reliability or validity data for this assessment.

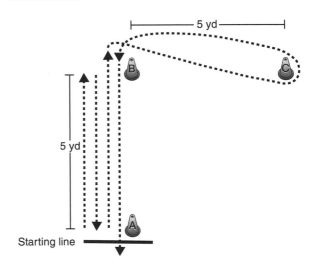

SLALOM TEST

▶ **Purpose:** To test multidirectional speed, agility, and body control.

▶ **Equipment:** Six hurdles, 60 cm high and 80 cm wide; standard stopwatch; 20 m open-space nonslip testing surface.

Procedure

1. Place six hurdles over a 12 m distance at 2 m intervals.
2. Instruct the client to run at maximum speed in a slalom pattern between the hurdles, both forward and backward to the starting point.
3. Measure from the time the first foot crosses the starting line to the time one of the feet crosses the same line for the last time.

Analysis and Interpretation of Data

Normative values for this test in 11-year-olds (males and females) are listed in table 10.12 on page 208.

Statistics

Reliability: ICC = 0.96 (Alricsson et al. 2001).

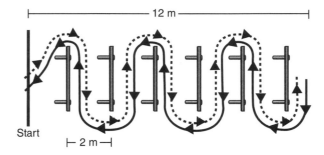

HURDLE TEST

▶ **Purpose:** To test multidirectional speed, agility, and body control.

▶ **Equipment:** Six hurdles, 60 cm high and 80 cm wide; standard stopwatch; 20 m open-space nonslip testing surface.

Procedure

1. Place the six hurdles over a 12 m distance at 2 m intervals (same course setup as for the slalom test).

2. Instruct the client to jump over the first hurdle, crawl under the second, jump over the third, and so on until he or she clears the sixth and final hurdle, then to sprint directly back to the starting line taking the shortest path.

3. Measure from the time the first foot crosses the starting line to the time one of the client's feet crosses the same line for the last time.

Analysis and Interpretation of Data

Normative values for this test among 11-year-old clients are listed in table 10.13 on page 208.

Statistics

Reliability: ICC = 0.90 (Alriccson et al. 2001).

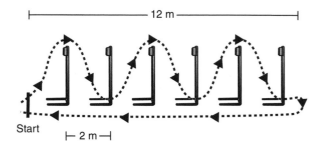

ILLINOIS AGILITY TEST

▶ **Purpose:** To test multidirectional speed, agility, and body control.

▶ **Equipment:** Flat nonslip surface, eight cones, stopwatch, measuring tape.

Procedure (Harman et al. 2000; Hoffman 2006)

1. Set up the course; length is 10 m and width is 5 m (distance between the start and finish points).
2. Use four cones to mark the start, the finish, and the two turning points.
3. Place another four cones down the center, spacing them 3.3 m apart.
4. Instruct the client to lie on his or her stomach (head to the start line) with hands by the shoulders.
5. Instruct the client on the "Go" command to get up as quickly as possible and run around the course in the direction indicated, without knocking the cones over, to the finish line.
6. Start timing when you issue the "Go" command and stop the stopwatch when the client reaches the finish line.

Analysis and Interpretation of Data

Table 10.14 on page 208 gives some rating scores for the test.

Statistics

There are no known reliability or validity data for this assessment.

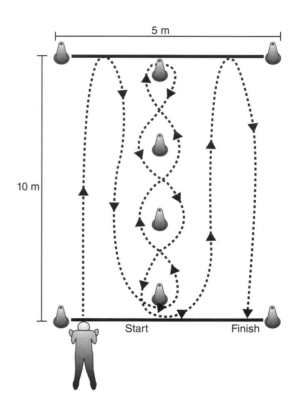

BACKWARD MOVEMENT AGILITY TEST

▶ **Purpose:** To test backward speed, agility, and body control.

▶ **Equipment:** Stopwatch, measuring tape, colored masking tape, tennis racket, timing lights (if available).

Procedure (Buckeridge et al. 2000)

1. Ask the client to hold a tennis racket in his or her playing hand.
2. Have the client assume the starting position on the starting mark (point A), facing the net.
3. Have the client hold the racket in the smash ready position.
4. Instruct the client to begin on command, and start the stopwatch at that time.
5. Instruct the client to move in a backward direction to the service line.
6. Allow the client to use either a sideways or a crossover step to move back to hit a smash through the finish gate (point B).
7. Have the client complete three timed trials, and record each one.
8. Record the best time as the final score.

Analysis and Interpretation of Data

Mean values with standard deviation and range for high-performance tennis players performing this test are listed in table 10.15 on page 208.

Statistics

There are no known reliability or validity data for this assessment.

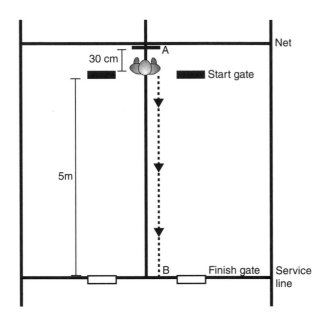

ZIGZAG RUN TEST

▶ **Purpose:** To test power, speed, quickness, and body control in multiple planes of movement. The test also assesses lower extremity control, including the ability to perform plant and cut types of movements correctly.

▶ **Equipment:** Standard measuring tape, two stopwatches, colored tape, five cones in the configuration shown in the figure.

Procedure (Ortiz et al. 2005)

1. Set up the path, which is 3 by 4.85 m and marked with colored tape on the floor, and place cones in every corner.
2. Position the client so that he or she is in a ready stance directly behind starting line.
3. Have the client run one lap as fast as possible around the zigzag path.
4. If the client does not round all cones in completing the one full lap, require him or her to rest for an adequate amount of time and then retest.
5. Record the time taken to perform one full lap around the path to the nearest tenth of a second.

Analysis and Interpretation of Data

■ Normal values (mean time ± SD) = 6.86 ± 0.53 (athletes; intertester testing) (n = 25) (Ortiz et al. 2005).

■ Normal values (mean time ± SD) = 7.67 ± 0.66 (nonathletes; intertester testing) (n = 25).

■ Normal values (mean time ± SD) = 6.97 ± 0.51 (athletes; intratester testing).

■ Normal values (mean time ± SD) = 7.70 ± 0.61 (nonathletes; intertester testing).

Statistics

■ Intertester (ICC; SEM) = 0.97 (0.18) in athletes; 0.97 (0.21) in nonathletes (Ortiz et al. 2005).

■ Intratester (ICC; SEM) = 0.92 (0.27) in athletes; 0.94 (0.30) in nonathletes (Ortiz et al. 2005).

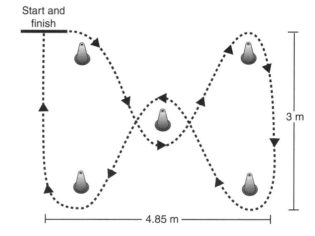

TABLE 10.1

T-Test Normative Values for Athletes

Sport	Time (s)
College basketball (men)	8.9
College basketball (women)	9.9
College baseball (men)	9.2
College tennis (men)	9.4
College tennis (women)	11.1
Competitive college athletes (men)	10.0
Competitive college athletes (women)	10.8
Recreational college athletes (men)	10.5
Recreational athletes (women)	12.5

Values are either means or 50th percentiles.

Adapted, by permission, from NSCA, 2008, Administration, scoring, and interpretation of selected tests, by E. Harman, J. Garhammer, and C. Pandorf. In *Essentials of strength training and conditioning*, 3rd ed., edited by T.R. Baechle and R.W. Earle (Champaign, IL: Human Kinetics), 278.

TABLE 10.2

Mean Scores (±SD) for the T-Test in Female and Male College-Age Subjects

Group	n	T-test norms (s)
FEMALE SUBJECTS		
Recreational sport	52	12.52 ± 0.90
College athletes	56	10.94 ± 0.60
MALE SUBJECTS		
Recreational sport	58	10.49 ± 0.89
College athletes	47	9.94 ± 0.50

Reprinted, by permission, from K. Pauole et al., 2000, "Reliability and validity of the T-test as a measure of agility, leg power, and leg speed in college-aged men and women," *Journal of Strength Conditioning Research* 14:443-450.

TABLE 10.3

T-Test Deciles for Female and Male College-Age Subjects (*n* = 152)

Deciles	Females T-test (s)	Males T-test (s)
90	10.69	9.45
80	10.92	9.82
70	11.48	10.06
60	11.99	10.22
50	12.34	10.44
40	12.73	10.64
30	13.24	10.99
20	13.69	11.22
10	14.23	11.69

Reprinted, by permission, from K. Pauole et al., 2000, "Reliability and validity of the T-test as a measure of agility, leg power, and leg speed in college-aged men and women," *Journal of Strength Conditioning Research* 14:443-450.

Percentile Ranks for the T-Test (s) in NCAA Division III College Football Players

% rank	Team	DB	DL	LB	OL	RB	WR	QB/TE
90	8.39	8.17	8.44	8.36	9.24	8.33	8.33	8.51
80	8.57	8.39	8.77	8.78	9.50	8.62	8.49	8.62
70	8.72	8.57	8.98	8.95	9.65	8.73	8.54	8.75
60	8.90	8.68	9.48	9.00	9.69	8.89	8.64	8.85
50	9.01	8.87	9.59	9.07	9.88	8.95	8.80	9.10
40	9.15	8.94	9.69	9.16	10.03	9.05	8.89	9.20
30	9.38	9.05	9.91	9.24	10.28	9.13	9.09	9.23
20	9.60	9.12	10.05	9.35	10.49	9.23	9.26	9.33
10	10.01	9.41	10.30	9.58	10.81	9.69	9.49	9.43
$\bar{X}$	9.11	8.81	9.45	9.05	9.93	8.95	8.85	8.99
SD	0.64	0.46	0.72	0.49	0.63	0.46	0.41	0.35
n	458	63	50	57	48	37	45	31

DB = defensive back; DL = defensive lineman; LB = linebacker; OL = offensive lineman; RB = running back; WR = wide receiver; QB = quarterback; TE = tight end.

Data collected using handheld stopwatches.

Reprinted, by permission, from J. Hoffman, 2006, *Norms for fitness, performance, and health* (Champaign, IL, Human Kinetics), 113.

Percentile Ranks for the T-Test (s) in Elite High School Soccer Players

% rank	Team
90	9.90
80	10.01
70	10.08
60	10.13
50	10.18
40	10.37
30	10.53
20	10.67
10	10.90
$\bar{X}$	10.30
SD	0.42
n	40

Reprinted, by permission, from J. Hoffman, 2006, *Norms for fitness, performance, and health* (Champaign, IL, Human Kinetics), 113.

TABLE 10.6

Descriptive Data (Mean ± SD) for T-Test in Various Athletic Populations

Source	Population	Gender	Mean ± SD (s)
BASKETBALL			
Latin et al. 1994	NCAA DI Guards Forwards Centers	M	8.95 ± 0.53 8.74 ± 0.41 8.94 ± 0.38 9.28 ± 0.81
RECREATIONAL			
Pauole et al. 2000	College aged	F M	12.52 ± 0.90 10.49 ± 0.89
SOCCER			
Vanderford et al. 2004	Elite youth U14 U15 U16	M	 11.6 ± 0.1 11.0 ± 0.2 11.7 ± 0.1
VOLLEYBALL			
Fry et al. 1991	NCAA DI	F	11.16 ± 0.38

Adapted, by permission, from J. Hoffman, 2006, *Norms for fitness, performance, and health* (Champaign, IL, Human Kinetics), 114-115.

TABLE 10.7

Percentile Ranks for the Pro Agility Test (s) in NCAA Division I College Athletes

% rank	Women's volleyball	Women's basketball	Women's softball	Men's basketball	Men's baseball	Men's football
90	4.75	4.65	4.88	4.22	4.25	4.21
80	4.84	4.82	4.96	4.29	4.36	4.31
70	4.91	4.86	5.03	4.35	4.41	4.38
60	4.98	4.94	5.10	4.39	4.46	4.44
50	5.01	5.06	5.17	4.41	4.50	4.52
40	5.08	5.10	5.24	4.44	4.55	4.59
30	5.17	5.14	5.33	4.48	4.61	4.66
20	5.23	5.23	5.40	4.51	4.69	4.76
10	5.32	5.36	5.55	4.61	4.76	4.89
$\bar{X}$	5.03	5.02	5.19	4.41	4.53	4.54
SD	0.20	0.26	0.26	0.18	0.23	0.27
n	81	128	118	97	165	869

Data collected using electronic timing device.

Reprinted, by permission, from J. Hoffman, 2006, *Norms for fitness, performance, and health* (Champaign, IL, Human Kinetics), 113.

TABLE 10.8

Pro Agility Times for College Football Players Participating in the NFL Combine

% rank	PRO AGILITY (S)							
	DL	LB	DB	OL	QB	RB	TE	WR
90	4.22	4.07	3.89	4.45	4.07	4.02	4.18	3.97
80	4.32	4.13	3.96	4.53	4.12	4.14	4.21	4.03
70	4.38	4.16	4.05	4.57	4.16	4.18	4.26	4.07
60	4.41	4.21	4.07	4.61	4.20	4.22	4.31	4.10
50	4.46	4.24	4.12	4.69	4.25	4.25	4.35	4.15
40	4.52	4.28	4.18	4.77	4.33	4.31	4.39	4.20
30	4.58	4.31	4.19	4.83	4.36	4.34	4.42	4.24
20	4.68	4.41	4.21	4.93	4.38	4.38	4.46	4.26
10	4.75	4.53	4.27	5.06	4.41	4.49	4.56	4.33
$\bar{X}$	4.48	4.26	4.11	4.74	4.26	4.26	4.35	4.15
SD	0.22	0.17	0.15	0.39	0.15	0.16	0.13	0.15
n	89	38	76	125	38	58	39	85

Data collected from 1999 NFL combine.

DL = defensive lineman; LB = linebacker; DB = defensive back; OL = offensive lineman; QB = quarterback; RB = running back; TE = tight end; WR = wide receiver.

Reprinted, by permission, from J. Hoffman, 2006, *Norms for fitness, performance, and health* (Champaign, IL, Human Kinetics), 114.

TABLE 10.9

Descriptive Data (Mean ± SD) for the Pro Agility Test in Various Athletic Populations

Source	Population	Gender	Mean ± SD
FOOTBALL			
Wroble & Moxley 2001	High school	Male	5.02 ± 0.24

Sawyer et al. 2002	NCAA Division I
	    OL, DL
	    WR, DB
	    RB, TE, LB
		Male	4.53 ± 0.22
			4.35 ± 0.11
			4.35 ± 0.12
			4.6 ± 0.2

Stuempfle et al. 2003	NCAA Division II
	    OL
	    DL
	    OB
	    DB
		Male	4.6 ± 0.2
			4.8 ± 0.2
			4.8 ± 0.2
			4.5 ± 0.2
			4.6 ± 0.2

McGee & Burkett 2003	NFL drafted rookies
	    Rounds 1 + 2
	    Rounds 6 + 7
		Male	
			4.38 ± 0.29
			4.45 ± 0.29

SOCCER			
Unpublished data	NCAA Division III	Female Male	4.88 ± 0.18 4.43 ± 0.17
VOLLEYBALL			
Unpublished data	NCAA Division III	Female	4.75 ± 0.19

DL = defensive lineman; LB = linebacker; DB = defensive back; OL = offensive lineman; QB = quarterback; RB = running back; TE = tight end; WR = wide receiver.

Reprinted, by permission, from J. Hoffman, 2006, *Norms for fitness, performance, and health* (Champaign, IL, Human Kinetics), 114-115.

TABLE 10.10

Normative Data for the 505 Agility Test (Australian Athletes)

Sport	Squad	n	RIGHT FOOT (S) Mean	SD	LEFT FOOT (S) Mean	SD
Basketball	Female ACT	7	2.59	0.16	2.56	0.18
	Male ACT	13	2.20	0.09	2.21	0.11
Hockey	Female ACT	10	2.51	0.12	2.48	0.10
	Male ACT	15	2.28	0.14	2.27	0.06
Tennis	Female AIS/VIS	12	2.38	0.08	2.43	0.09
	Male AIS/VIS	11	2.25	0.06	2.24	0.07

ACT = Australian Capital Territory; AIS = Australian Institute of Sport; VIS = Victorian Sports Institute.

Reprinted, by permission, from Australian Sports Commission, 2000, Protocols for the physiological assessment of team sport players, edited by L. Ellis et al. In *Physiological tests for elite athletes,* edited by C.J. Gore (Champaign, IL: Human Kinetics), 135.

TABLE 10.11

Three-Cone Drill Times for College Football Players Participating in the NFL Combine

% rank	THREE-CONE DRILL (S) DL	LB	DB	OL	QB	RB	TE	WR
90	7.22	7.05	6.87	7.66	7.06	7.17	7.12	6.85
80	7.45	7.16	6.97	7.82	7.13	7.29	7.16	7.01
70	7.52	7.30	7.07	7.98	7.19	7.32	7.27	7.10
60	7.64	7.38	7.09	8.07	7.31	7.36	7.38	7.19
50	7.71	7.49	7.14	8.15	7.36	7.47	7.42	7.28
40	7.78	7.54	7.22	8.28	7.40	7.53	7.48	7.35
30	7.89	7.61	7.29	8.38	7.54	7.60	7.57	7.41
20	8.07	7.70	7.39	8.51	7.59	7.71	7.71	7.49
10	8.47	7.84	7.47	8.66	7.70	7.82	8.04	7.58
$\bar{X}$	7.75	7.46	7.17	8.18	7.29	7.48	7.47	7.26
SD	0.43	0.30	0.22	0.43	0.57	0.27	0.34	0.30
n	88	57	102	139	38	58	41	86

Data collected from 1999 NFL combine.

DL = defensive lineman; LB = linebacker; DB = defensive back; OL = offensive lineman; QB = quarterback; RB = running back; TE = tight end; WR = wide receiver.

Reprinted, by permission, from J. Hoffman, 2006, *Norms for fitness, performance, and health* (Champaign, IL, Human Kinetics), 114.

TABLE 10.12

Normative Data for Slalom Test for 11-Year-Olds

	Mean (min-max) (s)	SD	CV
Males (n = 8)	12.28 (12.0-12.60)	.28	2.29
Females (n = 3)	11.90 (11.6-12.2)	.28	2.3

SD = standard deviation; CV = coefficient of variance.
Alricsson et al. 2001

TABLE 10.13

Normative Data for Hurdle Test for 11-Year-Olds

	Mean (min-max) (s)	SD	CV
Males (n = 8)	11.25 (10.73-11.85)	.52	4.48
Females (n = 3)	12.60 (11.93-13.6)	.78	5.97

SD = standard deviation; CV = coefficient of variance.
Alricsson et al. 2001

TABLE 10.14

Illinois Agility Test Ratings (s)

Rating	Male	Female
Excellent	<15.2	<17.0
Good	16.1-15.2	17.9-17.0
Average	18.1-16.2	21.7-18.0
Fair	18.3-18.2	23.0-21.8
Poor	>18.3	>23.0

Unpublished data.

TABLE 10.15

Backward Movement Agility Test Data for Female and Male High-Performance Tennis Players

Group	n	Time (s)
Female AIS scholarship holders	9	
Mean		1.43
SD		0.12
Range		1.26-1.63
Male AIS scholarship holders	3	
Mean		1.34
SD		0.12
Range		1.21-1.44
Female VIS scholarship holders	5	
Mean		1.49
SD		0.07
Range		1.42-1.58
Male VIS scholarship holders	8	
Mean		1.40
SD		0.11
Range		1.25-1.55

AIS = Australian Institute of Sport/national-level players; VIS = Victorian Institute of Sport/state-level players, SD = standard deviation.

Reprinted, by permission, from Australian Sports Commission, 2000, Protocols for the physiological assessment of high-performance tennis players, by A. Buckeridge et al. In *Physiological tests for elite athletes,* edited by C.J. Gore (Champaign, IL: Human Kinetics), 394.

part

III

Testing Procedures and Protocols for Regional Physical Parameters

Part III of the text describes tests for specific regions of the body: trunk (chapter 11), upper extremity (chapter 12), and lower extremity (chapter 13).

Research on lower extremity anaerobic testing has been reported in the literature for many years and has been conducted by investigators in multiple disciplines. Research on trunk testing is more recent, with increasing emphasis on "core" training because of evidence that lumbo-pelvic-hip weakness contributes to common dysfunctions and conditions such as anterior cruciate ligament tears, patellofemoral pain syndrome, ankle sprains, and

low back pain. Research on upper extremity testing, while less prevalent in the literature, has definite relevance to functional ability in the context of sport, exercise, and occupational activities.

As in part II, we outline the purpose of each test, describe how it is to be used, and present research findings when available. This part of the book is also organized from least to most complex with respect to the type of physical parameter and the tests. Again we encourage clinicians to read about each test and make deliberate decisions about which tests are appropriate for each of their clients.

Trunk Testing

Low back pain (LBP) will affect 60% to 80% of the population at some point in their lifetime (Greene et al. 2002). Among the many causes of LBP are spinal stenosis and spinal instability (Visuri et al. 2005; Jager et al. 2003). Potential contributing factors of spinal instability include lack of dynamic muscle control (Richardson et al. 1999; Jager et al. 2003; McGill 1998). Many of the lumbar spine and lower extremity injuries in athletes can also be attributed to muscular deficiencies such as weakness (Biering-Sorensen 1984; Beckman & Buchanan 1995; Nadler et al. 2000) and poor endurance (Biering-Sorensen 1984; McGill 1998; Udermann et al. 2003). It has been shown that trunk endurance has greater value than strength in preventing LBP (Alaranta et al. 1995; McGill et al. 2003). Endurance of the trunk extensors specifically predicts the occurrence of LBP among 30- to 60-year-old adults (Biering-Sorensen 1984). McGill and colleagues (2003) also demonstrated that clients with LBP have less endurance in their back extensors than in the flexors. This muscle imbalance is also a major focus of concern and should

be critically looked at during assessment of trunk endurance (McGill et al. 1999). Monitoring and correcting for muscle imbalances can potentially avoid future musculoskeletal injury and therefore time away from the job or playing field.

Trunk muscle activity occurs prior to activity in the lower extremities (Hodges & Richardson 1997). Preactivation of the trunk musculature in such a "feed-forward" mechanism is thought to stiffen the spine to provide a foundation for functional movements (Hodges & Richardson 1997), although the specific role of the inner muscle groups (transverse abdominis, multifidus, etc.), specifically the multifidus, has recently come under question (MacDonald et al. 2006). Further research in this area is probably prudent, but there does appear to be enough evidence to support the necessity of assessing for trunk muscle endurance and for balance between different muscle groups.

Neck pain is another common condition that has substantial personal and financial costs (Cote et al. 1998; Holmstrom et al. 1992). It has been demonstrated (Vernon et al. 1992) that clients with both whiplash-associated disorder (WAD) and insidious onset of neck pain have less strength in the cervical flexor muscles than asymptomatic clients. There was a progressive anterior-to-posterior muscle imbalance in the neck pain subjects, with the cervical flexors becoming relatively weaker as compared to the extensors. Jull and colleagues (2004) determined that altered patterns in muscle coordination within the neck flexor synergy were present in patients with neck pain of WAD and insidious onset. Not only has deep neck flexor muscle weakness been shown in clients with previous pathology; it has also been suggested as a causative or contributory factor in the pathogenesis of head and neck pain in general (Placzek et al. 1999; Watson & Trott 1993).

With this understanding of the exorbitant costs and the demonstrated muscle dysfunction contribution to these costs, it is the responsibility of the clinician to use appropriate testing to determine the presence or absence of such dysfunctions. This chapter presents the available tests or assessments appropriate for such determinations. We start with deep neck flexor muscle assessments and move distally to the various trunk dynamic and endurance assessments, concluding with standing assessments of the client's readiness for lifting and determination of lifting capacity. Each test is described in detail, and where appropriate the available statistical information is listed. As with all the testing covered in this book, we recommend that clinicians have a planned testing battery as discussed in chapters 2 and 3.

Assessment of Deep Neck Flexor Muscle Endurance

DEEP NECK FLEXOR TEST

▶ **Purpose:** To assess the synergistic muscle function of the various deep neck flexors (see table 11.1 on p. 233) and the sternocleidomastoid muscle.

▶ **Equipment:** Stable testing surface, standard stopwatch, nonpermanent marker.

Procedure

1. Position the client in the supine hook-lying position with knees bent 90° and hands on the abdomen.

2. Have the client maximally retract the chin and maintain this position isometrically, then lift the head and neck until the head is approximately 2.5 cm above the testing surface while the chin is kept retracted to the chest.

3. While the client maintains this position, draw a line across two approximated skinfolds along the client's neck and place your left hand on the table just below the occipital bone of the client's head.

4. Give verbal commands ("Tuck your chin" or "Hold your head up") when either the line edges begin to separate or the client's head touches your left hand.

5. Terminate the test if the edges of the lines no longer approximate each other or if the client's head touches your hand for more than 1 s.

6. Measure the time, to the nearest second, from initiation of the start position to test termination.

Analysis and Interpretation of Data

■ Mean endurance:
- For women without neck pain ($n = 38$), 14.5 s (SD 4.3); for men without neck pain ($n = 55$), 18.2 s (SD 3.3) (Grimmer 1994).
- For men and women without neck pain ($n = 20$), 37.1 to 44 s (SD 25.6-36.7); for men and women with neck pain ($n = 20$), 20.9 to 25.5 s (11-12.5) (Harris et al. 2005).
- For men and women with mechanical neck pain ($n = 22$), 5 s (SD 4).

Statistics

■ Intrarater reliability:
- Intraclass correlation coefficient (ICC) (3,1) = .82 to .91; SEM = 8.0 to 11.0; coefficient variation of method error (CVme) = 17.6 to 31 for clients without neck pain (Harris et al. 2005).
- ICC = .92 for women and .93 for men without neck pain (Grimmer 1994).

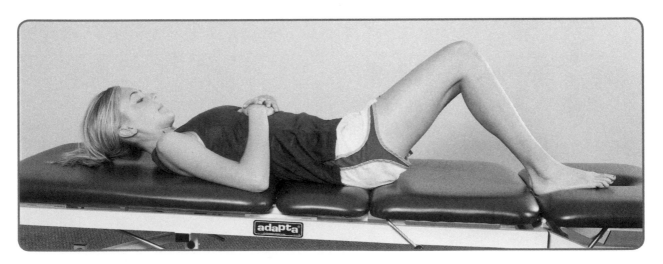

(continued)

(continued from previous page)

▪ Interrater reliability:
- ICC (2,1) = .67 to .81; SEM = 12.6 to 15.3; CVme = 25.0 to 36.1 for clients without neck pain (Harris et al. 2005).
- ICC (2,1) = .67; SEM = 11.5; CVme = 27.2 for clients with neck pain (Harris et al. 2005).
- ICC (2,1) = .57 (.14-.81); SEM = 2.3 for clients (n = 22) with mechanical neck pain (Cleland et al. 2006).

Note

This test position has been shown to maximally activate the deep neck flexor muscles (Vitti et al. 1973). A similar test (craniocervical flexion test) has been described in which the client starts in the same position. The subject performs five increments of increasingly inner-range craniocervical flexion in supine (Falla et al. 2003; Jull 2000), with the pressure initially inflated to 20 mmHg and gradually progressed (in 2 mmHg increments) to 30 mmHg with a 5 s hold at each stage (if successful) and a 10 s rest between stages. Clients are guided to the test level by feedback from a pressure unit (Stabilizer, Chattanooga, TN), which is placed behind the neck to monitor the progressive flattening of the cervical lordosis resulting from the contraction of longus colli (Mayoux-Benhamou et al. 1994, 1997). Performance in the test has been examined in subjects with WAD (Jull 2000) and cervicogenic headache (Jull et al. 1999). The results indicated that patients were less able to achieve and hold the progressive positions of the test than control subjects. These results implied dysfunction in the deep neck flexors. Similar results were obtained in a study comparing onset of neck pain from insidious-onset versus WAD pain and a control group, indicating that this physical impairment in neck flexor synergy is common to neck pain of both whiplash and insidious origin (Jull et al. 2004).

Assessment of Lower Trunk Dynamic Stability

SEGMENTAL MULTIFIDUS TEST

▶ **Purpose:** To assess the multifidus muscle's ability to provide segmental control at different lumbar levels, as well as comparisons from side to side. For the origin, insertion, action, and innervation of this muscle, see table 11.2 on page 233.

▶ **Equipment:** Solid, stable testing surface.

Procedure (Keely 2001)

1. Position the client so that he or she is hook-lying on a stable surface.
2. Stand directly at the side to be tested.
3. During the test, have the client perform transverse abdominis contraction, relaxing between changes of positions.
4. Palpate the L5-S1 interspace with your cranial hand and flex the lower extremities and trunk with your caudal hand until the L5-S1 interspace is initiated.
5. Then instruct the client "Hold, don't let me move you" while you provide a gentle, gradual lateral pressure to distal femurs, creating a rotational force through the lumbar spine.
6. Test both sides at each segmental level.
7. To assess higher lumbar segmental levels, passively flex bilateral lower extremities further until the corresponding segmental level is initiated.

Analysis and Interpretation of Data

■ Clinician compares right to left, as well as segmental level to segmental level.

■ Assessment is for lack of resistance to rotational force, that is, segmental weakness.

■ Increasing trunk flexion is required to assess more cranial levels.

Statistics

There are no known reliability or validity data for this assessment.

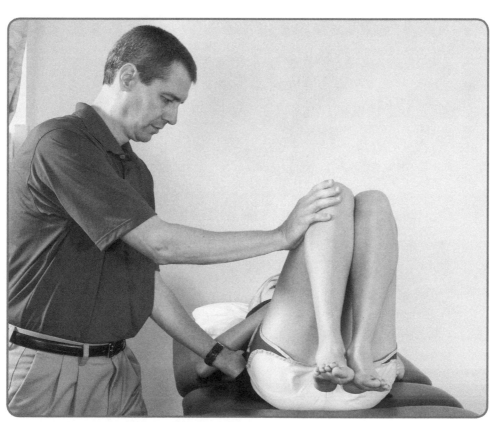

TRUNK CURL-UP TEST

▶ **Purpose:** To assess agonistic muscle interaction between iliopsoas and abdominal muscles with a trunk curl-up.

▶ **Equipment:** Stable testing surface.

Procedure (Janda 1983)

1. Position the client in supine with hips and knees flexed, both feet flat on the testing surface.
2. Have the client attempt to sit up from the supine position.
3. Note that little flexion of the trunk will be evident if the iliopsoas is dominant, as most of the flexion occurs at the hip.
4. Ask the client to perform a sit-up while actively plantarflexing the ankles, thus removing the effect of the iliopsoas.
5. Have the client progressively flex the spine, starting at the cervical spine, until the lumbar region is flexed.

Analysis and Interpretation of Data

■ The client's feet will lift from the surface as soon as iliopsoas becomes involved in the movement.
■ The client should be able to normally curl up so that the thoracic and lumbar spines are clear of the testing surface before the feet lift.
■ A client in excellent condition can complete a full sit-up without having the feet lift off the testing surface.

Statistics

There are no known reliability or validity data for this assessment.

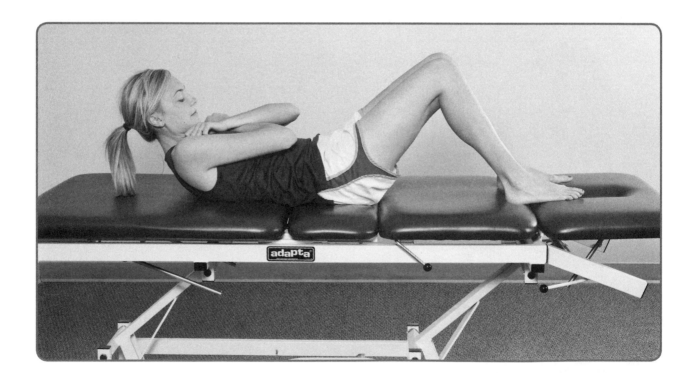

DOUBLE-LEG LOWERING TEST

▶ **Purpose:** To assess the ability of the trunk abdominal muscles to function eccentrically and control the lumbopelvic complex, and to assess pelvic tilt motor control.

▶ **Equipment:** Stable testing surface, goniometer.

Procedure

1. Position the client in supine; place both lower extremities in a vertical position.

2. Visually observe to ensure that the client's anterior superior iliac spines are in the same plane as the pubic symphysis.

3. Palpate the posterior pelvis and lower back area to monitor for tilting of the pelvis.

4. Instruct the client to perform an abdominal contraction with a posterior tilt of the pelvis in order to keep the lumbar spine flat throughout the test and prevent unwanted anterior tilting of the pelvis.

5. As the client lowers bilateral lower extremities to the treatment table while keeping the knees straight, monitor for anterior pelvis tilting.

6. You can also perform this test with a pressure cuff.

   • Have the client in the same starting position as already described with the pressure cuff placed under the lumbar spine at the level of L4 to L5.

   • Inflate the cuff to 40 mmHg and raise the client's legs until the pelvis posteriorly rotates and the needle on the pressure monitor begins to move.

   • Have the client perform the test as already described.

   • Measure the hip angle, as already explained, when the sphygmomanometer shows a 10 mmHg decrease in pressure (Richardson et al. 1999).

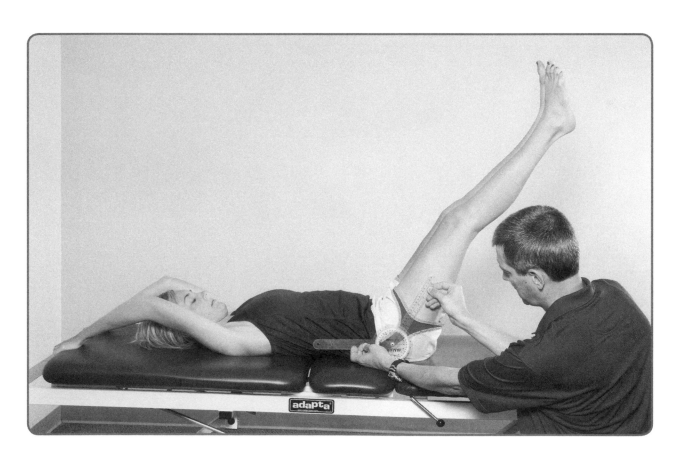

(continued)

(continued from previous page)

Analysis and Interpretation of Data

■ The angle from vertical to bilateral lower extremities is measured when it is determined that the client's pelvis is tilting anteriorly.

■ Muscle strength is assigned based on a previously established grading scale (table 11.3 on p. 233).

■ The average values in degrees for the double-leg lowering (DLL) test measured via blood pressure cuff monitoring are listed in table 11.4 on page 234.

Statistics

■ In a study demonstrating excellent test–retest reliability (ICC = 0.98), a significant difference was found between men and women on performance of the DLL test with a convenience sample of 100 healthy volunteers (50 men, 50 women; age range 18-29 years) (Krause et al. 2005).

■ Although electromyographic activity of the abdominal muscles has been shown to be substantial during the raising and lowering of the lower limbs in this test (Andersson et al. 1998; Bankoff & Furlani 1984; Flint 1965; Guimaraes et al. 1991; Gilleard & Brown 1994; Shields & Heiss 1997; Wickenden et al. 1992), the test has been viewed as a subjective test with minimal validity or reliability between testers (Cutter & Kevorkian 1999; Schmidt & Blanpied 1987; Schmidt et al. 1987; Smith et al. 1996).

■ The scoring system associated with the DLL test has come under question because healthy young subjects appeared not to be able to prevent pelvic tilting with the test. The pelvis tilted anteriorly (as measured by motion and data analysis) by the time the extremities were lowered a mean of 3.6° (Zannotti et al. 2002).

■ Reliability (ICC [2,1]) = .63; SEM = ± 6.0° using a blood pressure cuff (Lanning et al. 2006).

■ Ladeira and colleagues (2005) found that the reliability of the DLL test was very high ($r = 0.932$) but that the validity was low ($r = -0.338$ to -0.446), concluding that it had low validity for assessing abdominal strength. These authors felt that the test may be a useful tool to assess pelvic tilt motor control for spine stability but that it is not suitable for assessing muscle strength.

ABDOMINAL STAGE TEST

▶ **Purpose:** To assess trunk abdominal control in a multistaged test.

▶ **Equipment:** Stable testing surface, 2.5 kg (5.5 lb) and 5 kg (11 lb) weight plates.

Procedure

1. Position the client so that he or she is hook-lying for all starting positions.

2. Have the feet a comfortable distance apart (they are not to be held down by a partner or strap, etc.).

3. Complete the seven stages of the test in succession.

   • Stage 1: Palms over the knees: arms straight with hands resting on thighs. Client moves forward until the fingers are touching the patella.

   • Stage 2: Elbows over knees: arms straight with hands resting on thighs. Client moves forward until the elbows are touching the patella.

   • Stage 3: Forearms to thighs: arms crossed, in contact with the abdomen, and hands gripping opposite elbow. Client moves forward until the forearms touch midthighs.

   • Stage 4: Elbows to midthighs: arms crossed, in contact with the abdomen, and hands gripping the opposite shoulder. Client moves forward until elbows touch the midthighs.

   • Stage 5: Chest to thighs: arms behind head with hands gripping opposite shoulders. Client moves forward until the chest touches the thighs.

   • Stage 6: Chest to thighs with 2.5 kg mass: arms bent behind head with hands crossed and holding 2.5 kg mass. Client moves forward until the chest touches the thighs.

   • Stage 7: Chest to thighs with 5 kg mass: arms bent behind head with the hands crossed and holding 5 kg mass. Client moves forward until chest touches the thighs.

4. Allow the client up to three attempts to pass each stage.

Analysis and Interpretation of Data

*The client's score is the last stage completed successfully. Staging is as follows:

Level	Rating	Level	Rating
0	Very poor	4	Good
1	Poor	5	Very good
2	Fair	6	Excellent
3	Average	7	Elite

*The following are reasons for scoring the attempt unsuccessful: Either foot is lifted partially or totally off the surface; the arms or head is thrown forward in a jerky manner; the arms are moved from the nominated position; the hips are lifted off the surface; or the client is unable to maintain a 90° angle at the knee or is unable to complete the nominated sit-up.

*Normative values for soccer players are listed in table 11.5 on page 234.

Statistics

There are no known reliability or validity data for this assessment.

SIT-UP ENDURANCE TESTING

▶ **Purpose:** To assess trunk abdominal endurance with a dynamic trunk flexion activity over a 1 min period.

▶ **Equipment:** Stable testing surface, repetition counter (optional).

Procedure

1. Position the client in supine with shoulders and head flat on the testing surface.
2. Have the client flex bilateral hips to approximately 45° and knees to approximately 90°.
3. Have the client hold bilateral arms crossed over the chest.
4. At the start of the test, instruct the client to flex the neck toward the chest; the upper trunk is curled toward bilateral thighs until the upper torso is off the surface.
5. Allow the torso to flatten back onto the testing surface.
6. Throughout testing, ensure that bilateral feet, buttocks, and lower torso are held flat against the testing surface and that bilateral arms are held stationary.
7. Instruct the client to perform this sequence for as many repetitions as possible in a 1 min period.
8. Record the number of correct repetitions performed in 1 min. Do not count a repetition if the client fails to reach the up position, fails to maintain the initial testing position, or raises the buttocks off the testing surface.

Analysis and Interpretation of Data

Percentile ranking for this test is listed in table 11.6 on page 234.

Statistics

There are no known reliability or validity data for this assessment.

Assessment of Trunk Muscle Endurance

ENDURANCE OF LATERAL FLEXORS (SIDE BRIDGE)

▶ **Purpose:** To determine endurance of trunk lateral flexor muscles for side-to-side comparison, as well as in comparison ratios with other trunk endurance muscles covered in this chapter.

▶ **Equipment:** Stable testing surface (preferably mat table that can have a belt wrap underneath), standard stopwatch.

Procedure

1. Position the client in right side-lying with the right elbow propped on the table.
2. Have the client extend bilateral legs and place the top foot in front of the lower foot on the table for support.
3. Have the client lift the bilateral hips to align the back, hip, and lower extremities.
4. Instruct the client to support him- or herself, with the hips lifted to maintain a straight line over the full body length, on one elbow and the feet.
5. Have the client hold the uninvolved arm across the chest with the hand of that arm placed on the opposite shoulder.
6. Manually record the endurance time in seconds from the point at which the client assumed the testing position until the hips returned to the table.

Analysis and Interpretation of Data

Grading is as follows:

Normal (5)	Client able to lift pelvis off table and hold spine straight for 20 to 30 s
Good (4)	Client able to lift pelvis off table but has difficulty holding spine straight for 15 to 20 s
Fair (3)	Client able to lift pelvis off table but has difficulty holding spine straight for 10 to 15 s
Poor (2)	Client able to lift pelvis off table but cannot hold spine straight for 1 to 10 s
Trace (1)	Client unable to lift pelvis off table

Statistics

■ Test–retest reliability (right side bridge) ICC = 0.96 over five consecutive days.

■ Test–retest reliability (left side bridge) ICC = 0.99 over five consecutive days.

■ [Mean age (normals) 21 years (men: $n = 92$; women: $n = 137$)] (data from McGill et al. 1999).

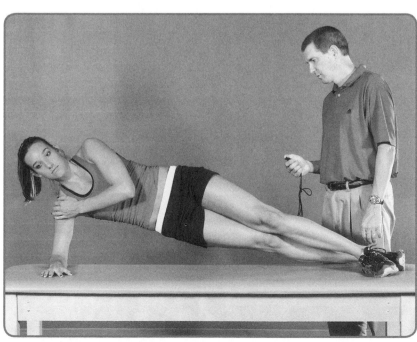

ENDURANCE OF EXTENSORS

▶ **Purpose:** To determine endurance of trunk extensor muscles in comparison ratios with other trunk endurance muscles covered in the chapter. Trunk endurance has been implicated in future LBP.

▶ **Equipment:** Stable testing surface (preferably mat table that can have a belt wrap underneath), standard stopwatch, two belts, stool.

Procedure

1. Position the client so that he or she is prone and the lower body is fixed with straps to the test surface at the ankles, knees, and hips as shown.

2. Have the upper body extended in a cantilevered fashion over the edge of the test surface and the arms resting on a stool (first position).

3. For testing, have the client cross the arms over the chest, with hands on opposite shoulders.

4. Instruct the client to lift the upper body off the stool until the upper torso is horizontal to the test surface (second position).

5. Instruct the client to maintain the horizontal position as long as possible.

6. Manually record the endurance time in seconds from the point at which the subject assumed the horizontal position until the upper body broke the horizontal plane.

Analysis and Interpretation of Data

Normative values, normative endurance ratios, and comparisons of normative values among various studies are listed in tables 11.7 through 11.10 on pages 235 through 236.

Statistics

■ Test–retest reliability ICC = 0.99 over five consecutive days.

■ [Mean age (normals) 21 years (men: $n = 92$; women: $n = 137$)] (data from McGill et al. 1999).

■ An alternative version, in which the clinician (as long as he or she is heavier in mass than the client) provides the stabilization (via sitting on client's feet) in place of belts, was also found to be reliable ($r = 0.90$) when compared with the original version described here (Reiman et al. 2008).

Notes

■ An alternative version of this assessment involves the use of a Roman chair. This alternative test has proven to be reliable in normal college-aged individuals (ICC; $r = 0.91$-0.96, $p < .05$) (Udermann et al. 2003).

■ Maximum isometric trunk extensor strength (not endurance) had no relationship to the presence of LBP, although a significant association between neuromuscular imbalance of right and left erector spinae and the occurrence of LBP was observed (Renkawitz et al. 2006).

■ Individuals with LBP have decreased muscle endurance in erector spinae muscles, and these muscles gradually develop anaerobic characteristics (Ashmen et al. 1996; Biering-Sorenson 1989; Bullock-Saxton 1994; Gracovetsky & Farfan 1986).

■ A stronger correlation has been found between extensor endurance and LBP than between flexor endurance and LBP (Alaranta et al. 1994, 1995).

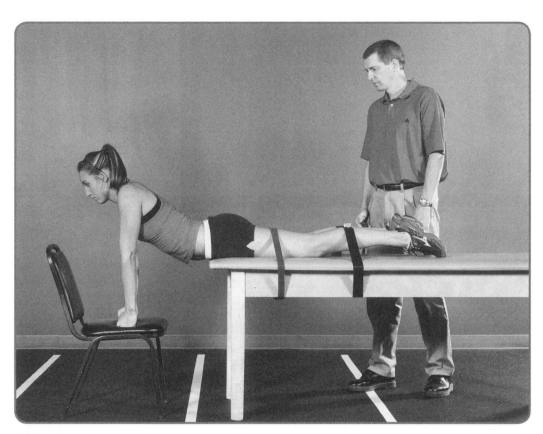

First position.

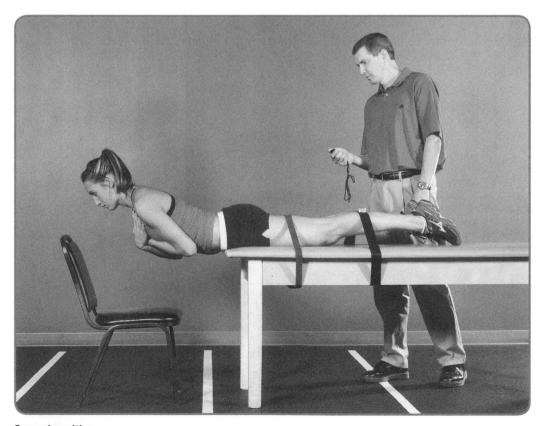

Second position.

ENDURANCE OF ABDOMINAL MUSCLES TEST

▶ **Purpose:** To determine endurance of trunk abdominal muscles in comparison ratios with other trunk endurance muscles covered in the chapter.

▶ **Equipment:** Stable testing surface that can be raised and lowered at one end (or a prefabricated 60° wedge if table cannot be adjusted), standard stopwatch, belt.

Procedure

1. Position the client so that he or she is sitting on the test surface, and place the upper body against a support with an angle of 60° from the test surface.

2. You can use a prefabricated wedge or use a plinth table with the angle at 60°.

3. Strap the client's feet to the test surface with a belt as shown.

4. Have the client's arms folded across the chest with the hands placed on the opposite shoulders (first position).

5. Instruct the client to maintain this body position while you remove the support (second position).

6. Manually record the endurance time in seconds from the point at which the client assumes the 60° position until the upper body falls below the 60° angle.

Analysis and Interpretation of Data

Normative values, normative endurance ratios, and comparisons of normative values among various studies are listed in tables 11.7 through 11.10 on pages 235 through 236.

Statistics

▦ Test–retest reliability ICC = 0.93 over five consecutive days.

▦ [Mean age (normals) 21 years (men: $n = 92$; women: $n = 137$)] (data from McGill et al. 1999).

▦ An alternative version, with the clinician (as long as he or she is heavier in mass than the client) providing the stabilization (by lying over the middle of bilateral lower extremities) in place of belts, was also found to be reliable ($r = 0.84$) in comparison to the original version described here (Reiman et al. 2008).

Notes

▦ Alternative testing positions for abdominal endurance have been advocated (Chen et al. 2003). Testing in the same manner as McGill (1999) but using the initial starting position (with a wedge) of 45° and a trunk curl-up position (with bilateral scapulae clearing the table) was more time effective and showed less variance than testing with the 60° trunk starting position (Chen et al. 2003). The curl-up exercise was considered easy, convenient, and representative of trunk flexor effort; it was considered a preferable alternative to the 60° flexor exercise for healthy females ($n = 28$) of mean age 23.8 ± 2.4 years (Chen et al. 2003).

▦ Testing the balance of endurance among trunk flexors, extensors, and lateral flexors better discriminates those with back pain or dysfunction than those who do not have back pain and/or dysfunction. (McGill et al. 1999). The issue may not necessarily be only extensors or flexors and so on, but rather the ratio among the various muscle groups. Neuromuscular imbalance rather than isolated strength deficits has been proven to be associated with LBP (Renkawitz et al. 2006); therefore it would stand to reason that endurance imbalances would also demonstrate such associations.

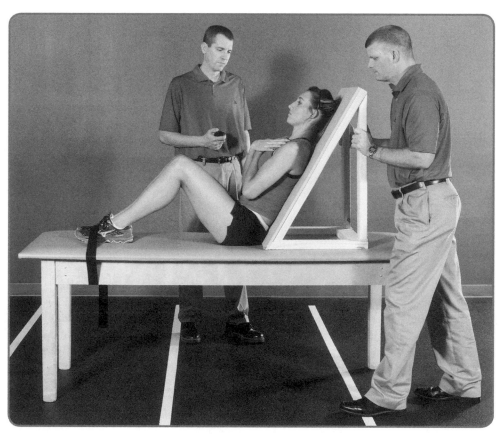
First position.

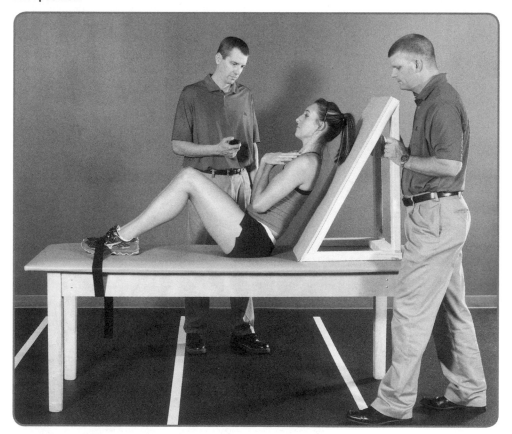
Second position.

ABDOMINAL DYNAMIC ENDURANCE

▶ **Purpose:** To assess dynamic abdominal muscle endurance in the supine position at a specified repetition rate.

▶ **Equipment:** Stopwatch, solid table for testing surface, two strips of tape across testing surface, prerecording (cassette, etc.) with 25 repetitions per minute beat.

Procedure

1. To start, have the client in a hook-lying position with the knees at 90°, and have him or her tuck the chin in as shown in the figure.
2. Place lines of tape at 8 cm (for clients 40 years of age or more) and 12 cm (for clients less than 40 years) from the third fingertip.
3. Instruct the client to perform a trunk curl to touch the age-appropriate line at a cadence of 25 repetitions per minute to a prerecorded repeated beat.
4. During the test you may correct technique or speed one time. Do not inform the client about the number of repetitions during the testing.
5. Record the number of repetitions completed.

Analysis and Interpretation of Data

■ Normal values = 31 repetitions ± 24 (abdominal dynamic endurance test with a maximum limit of 75 repetitions) (Moreland et al. 1997).
■ Females = 19 repetitions; males = 27 repetitions (Alaranta et al. 1994).

Statistics

■ ICC (2,1) = 0.89, SEM = 8 repetitions; 39 normal subjects with a mean age of 35 ± 9.3 (Moreland et al. 1997); 0.91 (Alaranta et al. 1994).
■ Mean values and reliability values for this assessment for various types of clients are listed in table 11.11 on page 237.
■ A significant negative correlation ($p < 0.001$) was noted between abdominal dynamic endurance testing scores and disability index scores (greater abdominal endurance scores correlated with decreased disability index scores and vice versa) among 89 females and 94 males (Rissanen et al. 1994).

Note

The repetitive arch-up test (as described in Alaranta et al. 1994) differs from the test described here in that the client's feet are fixed by the tester in the ankle region and the client moves the arms toward the knees, touching the kneecaps with the thenar region.

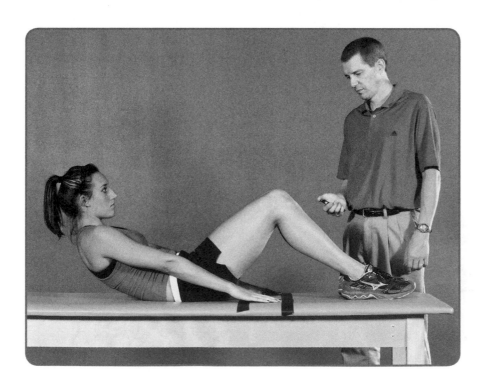

EXTENSOR DYNAMIC ENDURANCE TEST

▶ **Purpose:** To assess dynamic extensor endurance in a prone position at a specified repetition rate.

▶ **Equipment:** 30° foam wedge, stopwatch, two straps for support, solid table, prerecording (cassette, etc.) with 25 repetitions per minute beat.

Procedure

1. Position the client so that he or she is prone over a 30° foam wedge with the iliac crests at the edge of the wedge.

2. Use two straps to support the client's pelvis and lower extremities on the table, one at the hips and the other at the midcalf level.

3. Have the client hold the arms alongside the trunk with the hands at the hips.

4. Instruct the client to keep the spine straight and to extend the trunk to neutral and the lower back down so that the nose touches the table.

5. Instruct the client to perform this movement at a cadence of 25 repetitions per minute to a prerecorded beat.

6. During the test you may correct technique or speed one time. Do not inform the client about the number of repetitions during the testing.

7. Record the number of repetitions completed.

Analysis and Interpretation of Data

■ Normal values = 46 repetitions ± 18 (maximum repetitions) (Moreland et al. 1997).

■ Females = 24 repetitions; males = 28 repetitions (Alaranta et al. 1994).

Statistics

■ ICC (2,1) = 0.78, SEM = 9 repetitions; 39 normal subjects with a mean age of 35 ± 9.3 (Moreland et al. 1997); 0.83 (Alaranta et al. 1994).

■ Mean values and reliability values for this assessment for various types of clients are listed in table 11.12 on page 238.

■ A significant negative correlation ($p < 0.001$) was noted between extensor dynamic endurance testing scores and disability index scores (greater extensor endurance scores correlated with decreased disability index scores and vice versa) among 89 females and 95 males (Rissanen et al. 1994).

Note

The repetitive arch-up test (as described in Alaranta et al. 1994) differs from the test described here in that the client is held fixed by the tester in the ankle region and the client starts in the prone position with the upper trunk flexed over the edge of the table at a 45° angle, then moves the upper trunk up to the horizontal position and back down to 45° flexion.

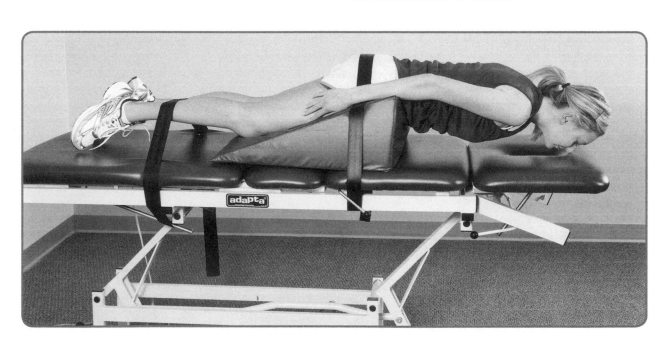

PRONE BRIDGE

▶ **Purpose:** To assess static trunk endurance and control in a prone position.

▶ **Equipment:** Stable testing surface, standard stopwatch.

Procedure

1. Position the client so that he or she is in the prone position, propped on the elbows; bilateral elbows are placed shoulder-width apart, and the feet are set with a narrow base but not touching.

2. Instruct the client to elevate the pelvis from the testing surface so that only the forearms and the toes are in contact with the surface.

3. Have the client maintain the shoulders, hips, and ankles in a straight line.

4. Instruct the client to hold this position until prevented from doing so by fatigue or pain.

5. Record the time in seconds from initiation of the testing position until the client is no longer able to maintain this position.

Analysis and Interpretation of Data

- Mean bridge durations for clients without back pain = 72.5 ± 32.6 s.

- Mean bridge durations for clients with back pain = 28.3 ± 26.8 s.

- Mean values (±SD) in asymptomatic and symptomatic clients for this assessment are listed in table 11.13 on page 239.

Statistics

Pearson correlation coefficients for test–retest reliability, $r = 0.78$ (Schellenberg et al. 2007).

SUPINE BRIDGE

▶ **Purpose:** To assess static trunk endurance and control in a supine position.

▶ **Equipment:** Stable testing surface, standard stopwatch.

Procedure

1. Position the client in supine with knees flexed 90° and the soles of the feet on the testing surface with a narrow base, but not touching; the hands are positioned by the ears.

2. Instruct the client to raise the pelvis from the testing surface so that the shoulders, hips, and knees are maintained in a straight line.

3. Have the client hold the position until prevented from doing so by fatigue or pain.

4. If the client reaches 2 min, have him or her extend the dominant leg at the knee, removing one point of support.

5. In the case of a preexisting injury to the dominant support leg, have the client extend the nondominant leg.

6. Record the time in seconds from initiation of the testing position until the client is no longer able to maintain the position.

Analysis and Interpretation of Data

■ Mean bridge durations for clients without back pain = 170.4 ± 42.5 s.

■ Mean bridge durations for clients with back pain = 76.7 ± 48.9 s.

■ Mean values (±SD) in asymptomatic and symptomatic clients for this assessment are listed in table 11.14 on page 239.

Statistics

Pearson correlation coefficients for test–retest reliability, $r = 0.84$ (Schellenberg et al. 2007).

Note

Extension of the dominant leg is designed to shorten the bridge duration by increasing the difficulty via addition of a torque moment to the core and an increase in the counterbalance weight.

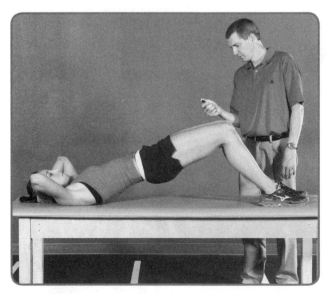

Bilateral legs.

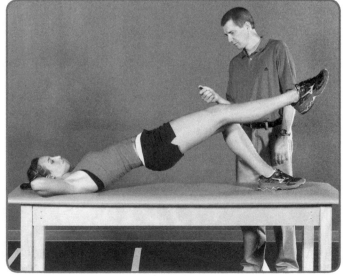

Single leg.

Assessment of Trunk Muscle Endurance and Lifting Capacity

LOADED FORWARD REACH

▶ **Purpose:** To assess trunk control and balance in a standing position with weight extended in front of the body.

▶ **Equipment:** A stick, wand, or dowel rod that is longer than client's shoulder width, standard tape measure, 4.5 kg (10 lb) weight.

Procedure

1. Have the client stand, holding the stick in both hands with a weight of 5% of his or her body weight (not exceeding 4.5 kg) hung from the stick and at shoulder height and width in both hands; bilateral elbows are extended forward at shoulder level.

2. Have the client reach as far forward as possible without lifting the heels off the floor.

3. Record the distance reached in centimeters.

Analysis and Interpretation of Data

No normative values listed for this test.

■ Mean (SD) = 50.62 (13.09) to 53.37 (13.27) cm; n = 52 patients with more than three months existing disability due to nonspecific LBP with a mean age of 43.19 ± 9.27 years, 25 of whom were male (Smeets et al. 2006).

■ Mean ± SD (n = 103 adults with LBP) = 53.78 ± 19.28 cm (Novy et al. 2002).

Statistics

ICC (1,1) = 0.74 (0.59-0.84) (Smeets et al. 2006).

PROGRESSIVE ISOINERTIAL LIFTING EVALUATION (PILE)

▶ **Purpose:** To assess the client's ability to perform repetitive lifting as quickly as possible. This assessment could help the clinician determine a client's readiness to return to any task requiring repetitive lifting.

▶ **Equipment:** Box to be lifted, 75 cm high table, weights of various proportions.

Procedure (Smeets et al. 2006)

1. Position the client directly in front of the box to be lifted.
2. Remind the client of proper posture and lifting mechanics prior to performance of the test.
3. Instruct the client to lift the box containing a weight four times within 20 s from the floor up to the 75 cm high table.
4. Record the number of fully completed lifting stages.

Analysis and Interpretation of Data

- In the original study the starting weight (including weight of box) was 3.6 kg (7.9 lb) for women and 5.85 kg (12.9 lb) for men. After each completed lifting cycle, the weight was increased by 2.25 kg (5 lb) for women and by 4.5 kg (10 lb) for men.
- Testing was completed when the client could not lift the box four times within 20 s, the client decided to stop because of fatigue or pain, the heart rate exceeded 85% of the maximal heart rate, the maximal amount of weight that could safely be lifted was reached (60% of body weight), or the clinician determined that the lifting was unsafe.
- Mean (SD) = 4.10 (2.61) to 4.27 (2.80) lifting stages; $n = 50$ patients with more than three months existing disability due to nonspecific LBP with a mean age of 43.19 ± 9.27 years, 25 of whom were male.

Statistics

ICC (1,1) = 0.92 (0.87-0.96).

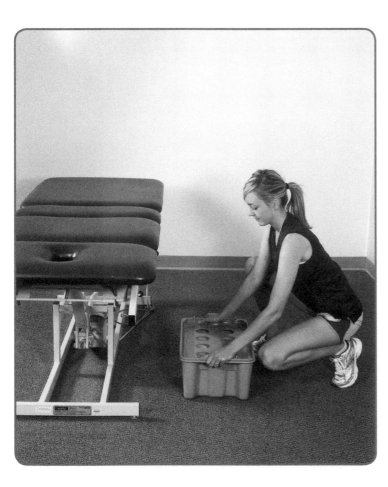
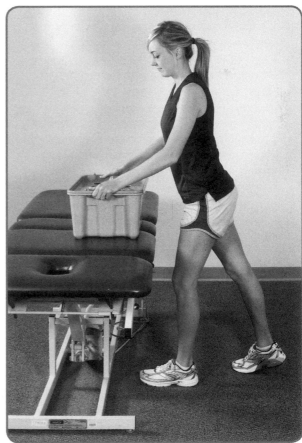

REPETITIVE BOX-LIFTING TASK (RBLT)

▶ **Purpose:** To assess the client's ability to repetitively lift a specified weight from the floor to a platform over 10 min. This assessment could help the clinician determine a client's readiness to return to any task requiring repetitive lifting.

▶ **Equipment:** 20.5 kg (45 lb) boxes (preferably metal) with side handles, two 1.3 m high platforms, standard stopwatch or electronic timing device. In addition to the equipment, two to four assistants or technicians and a scorekeeper are needed.

Procedure

1. Position the client so that he or she is standing directly in front of the box to be lifted.
2. Remind the client of proper posture and lifting mechanics prior to performance of the test.
3. Instruct the client to lift the 20.5 kg boxes from the floor onto a 1.3 m high platform continuously for 10 min.
4. Have the client move back and forth between the two platforms (2.4 m apart), each time lifting the box from the floor.
5. After each lift, assistants lower the box to the floor and position it for the next lift (the client lifts but never lowers the boxes).
6. Instruct the client to lift as many times as possible during the 10 min period.
7. Record the number of times the boxes were lifted in the 10 min period.

Analysis and Interpretation of Data

No normative values listed for this test.

Statistics

▪ ICC = 0.94 over four trials (Pandorf et al. 2003).
▪ ICC = 0.93 for three trials and 0.97 for two trials (Sharp et al. 1993).
▪ ICC = 0.97 (Knapik & Gerber 1996).

TABLE 11.1

Deep Neck Flexor Muscles

Muscle	Origin	Insertion	Action	Innervation
Rectus capitis anterior	C1 (atlas) transverse process and lateral mass	Occiput (basilar part, inferior surface)	Capital flexion	C1 & C2
Rectus capitis lateralis	C1 (atlas) transverse process	Occiput (jugular process)	Capital flexion	C1 & C2
Longus capitis	C3-C6 vertebrae transverse processes, anterior tubercles	Occiput (basilar part, inferior surface)	Capital flexion	C1-C3
Longus colli	C3-C5 transverse processes, T1-T3 and C5-C7 vertebral bodies (anterolateral bodies), T1-T3 anterior vertebral bodies	Anterior arch and tubercle of atlas, C2-C4 anterior vertebral bodies, C5-C6 transverse processes, anterior tubercles	Cervical and capital flexion	C2-C6

TABLE 11.2

Multifidus Muscle

Muscle	Origin	Insertion	Action	Innervation
Multifidus	At each level, several fascicles arise via a common tendon from the spinous process	Diverges caudally to assume separate attachments to the mamillary process, iliac crest, and sacrum; some of the deeper fibers attach to facet joint capsule	Opposes flexion effect of abdominal muscles as they produce rotation; stabilizes vertebrae	Dorsal rami of spinal nerves

TABLE 11.3

Double-Leg Lowering Test Muscle Grading

Angle (Cutter and Kevorkian 1999; Kendall 1993)	Muscle grade (nominal scale)	Muscle grade (numerical scale)	Muscle grade (Reese 1999)
0°	Poor	2	Able to reach 75° to 45° from the table before pelvis tilts
15°	Fair	3	Able to reach 46° to 90° from the table
30°	Fair plus	3+	
45°	Good minus	4–	
60°	Good	4	Able to reach 16° to 45° from the table
75°	Good plus	4+	
90°	Normal	5	Able to reach 0° to 15° from the table

TABLE 11.4

Double-Leg Lowering Test Average Raw Values (degrees)*

Sport	Sex	n	Mean	SD
Soccer	Women	19	50	11
	Men	35	48	11
Cross country	Women	14	48	10
	Men	12	44	9
Field hockey	Women	15	50	10
Volleyball	Women	10	59	7
Average	Women	58	52	5
	Men	47	46	3
Total average		105	50	10

* Measured with blood pressure cuff.

TABLE 11.5

Abdominal Stage Test Normative Values (Soccer)

Sport	Sex	n	Mean	SD	Range
Soccer	Female	23	4	1.1	1-5
Soccer	Male	37	3	1.3	1-6

Reprinted, by permission, from Australian Sports Commission, 2000, Protocols for the physiological assessment of team sport players, by L. Ellis et al. In *Physiological tests for elite athletes,* edited by C.J. Gore (Champaign, IL: Human Kinetics), 144.

TABLE 11.6

Interpretation of Sit-Up Endurance Test

Percentile rank	AGES 20-29		AGES 30-39	
	Men	Women	Men	Women
99	>55	>51	>51	>42
90	52	49	48	40
80	47	44	43	35
70	45	41	41	32
60	42	38	39	29
50	40	35	36	27
40	38	32	35	25
30	35	30	32	22
20	33	27	30	20
10	30	23	26	15
01	<27	<18	<23	<11

Reprinted, by permission, from NSCA, 2000, Administration, scoring, and interpretation of selected tests, by E. Harman, J. Garhammer, and C. Pandorf. In *Essentials of strength training and conditioning,* 2nd ed., edited by T.R. Baechle and R.W. Earle (Champaign, IL: Human Kinetics), 311.

TABLE 11.7

Endurance Ratios Among Muscle Groups Studied

Assessment	Men	Women	All
Flexion/extension	0.84	0.72	0.77
RSB/LSB	0.96	0.96	0.96
RSB/extension	0.58	0.40	0.48
LSB/extension	0.61	0.42	0.50

Mean age 21 years (men: $n = 92$; women: $n = 137$).
RSB = right side-bending; LSB = left side-bending.

TABLE 11.8

Endurance Ratio Approximations Among Muscle Groups Studied

Assessment	Approximations
Flexion/extension	0.75 (or ¾)
RSB/LSB	1.0 (1/1)
RSB/extension	0.50 (or ½)
LSB/extension	0.50 (or ½)

Mean age 21 years (men: $n = 92$; women: $n = 137$).
RSB = right side-bending; LSB = left side-bending.

TABLE 11.9

Comparison of Endurance Testing Normative Values

Assessment	Mean ± SD (s) (McGill et al. 1999)	Ratio	Mean ± SD (s) (Chan 2005)	Ratio	Mean ± SD (s) (Reiman et al. 2006)	Ratio
Extension	171 ± 60	1.0	114.28 ± 34.62	1.0	113.86 ± 45.59	1.0
Flexion	147 ± 90	0.90	176.56 ± 88.58	1.54	186 ± 77.26	1.63
Left side bend	85 ± 36	0.47	94.53 ± 32.97	0.83	59.77 ± 20.14	0.53
Right side bend	81 ± 36	0.50	98.13 ± 41.38	0.86	57.37 ± 20.38	0.50

TABLE 11.10

Comparison of Endurance Testing Values for Various Normal Subjects, Sports, and Injury Levels

Study	Extension (s)	Side bend (s)
McGill et al. 1999		
College-aged		
Normals: Male (n = 31)	146 ± 51	97 ± 35 (left)
		94 ± 34 (right)
Female (n = 44)	189 ± 60	77 ± 35 (left)
		72 ± 31 (right)
Chan 2005		
Normals: Male rowers (n = 42)	114.28 ± 34.62	94.53 ± 32.97 (left)
		98.13 ± 41.38 (right)
Reiman et al. 2006		
High school weightlifting athletes		
Normals: Male (n = 21)	108.94 ± 40.75	64.75 ± 20.62 (left)*
		62.73 ± 17.00 (right)*
Female (n = 17)	113.99 ± 54.05	49.11 ± 19.06 (left)
		46.34 ± 22.01 (right)
Leetun et al. 2004		
Normals: Male (n = 60)	130.4 ± 40.0	84.3 ± 32.5
Female (n = 79)	123.4 ± 48.4	58.9 ± 26.0
Sport		
Male basketball (n = 44)	131.4 ± 42.0	82.7 ± 30.6
Male cross country (n = 17)	122.9 ± 38.7	87.6 ± 37.1
Female basketball (n = 60)	115.7 ± 43.5	57.8 ± 24.7
Female cross country (n = 18)	151.4 ± 52.5	60.9 ± 30.5
Uninjured (n = 99)	128.3 ± 43.6	72.0 ± 32.4
Injured (n = 41)	121.6 ± 48.9	64.7 ± 28.8

*Significant difference in males versus females in left and right side bend (p < .05).

TABLE 11.11

Abdominal Dynamic Endurance Adjusted Mean Values (s) for Subgroups

Category	n	Mean	ICC (SEM)
Sex			
Female	24* 233#	20#	.94 (6)*
Male	15* 242#	27#	.82 (10)*
Occupation			
Office or sedentary job type	15*		.88 (9)*
Light or medium job type	13*		.92 (7)*
Heavy or very heavy job type	11*		.93 (7)*
White collar	181#	29#	
Blue collar	294#	21#	
Age			
35-39	123#	28#	
40-44	136#	24#	
45-49	107#	23#	
50-54	109#	19#	
Blue-collar female			
35-39		24 ± 12#	
40-44		18 ± 12#	
45-49		17 ± 14#	
50-54		9 ± 10#	
Blue-collar male			
35-39		29 ± 13#	
40-44		22 ± 11#	
45-49		19 ± 11#	
50-54		17 ± 13#	
White-collar female			
35-39		30 ± 16#	
40-44		19 ± 13#	
45-49		22 ± 15#	
50-54		20 ± 13#	
White-collar male			
35-39		35 ± 13#	
40-44		34 ± 12#	
45-49		33 ± 15#	
50-54		36 ± 16#	

ICC = Intraclasss Correlation Coefficient; SEM = Standard Error of Mean.

Data compiled from Alaranta et al. 1994 (#) and Moreland et al. 1997 (*).

TABLE 11.12

Extensor Dynamic Endurance Adjusted Mean Values (s) for Subgroups

Category	n	Mean	ICC (SEM)
Sex			
Female	24*		.73 (10)*
	233#	25#	
Male	15*		.82 (9)*
	242#	28#	
Occupation			
Office or sedentary job type	15*		.84 (8)*
Light or medium job type	13*		.73 (12)*
Heavy or very heavy job type	11*		.74 (7)*
White collar	181#	30#	
Blue collar	294#	24#	
Age			
35-39	123#	27#	
40-44	136#	26#	
45-49	107#	28#	
50-54	109#	25#	
Blue-collar female			
35-39		28 ± 13#	
40-44		25 ± 14#	
45-49		25 ± 15#	
50-54		18 ± 14#	
Blue-collar male			
35-39		26 ± 11#	
40-44		23 ± 12#	
45-49		24 ± 13#	
50-54		21 ± 11#	
White-collar female			
35-39		27 ± 11#	
40-44		20 ± 11#	
45-49		31 ± 16#	
50-54		26 ± 14#	

Category	n	Mean	ICC (SEM)
White-collar male			
35-39		34 ± 14#	
40-44		36 ± 14#	
45-49		34 ± 16#	
50-54		35 ± 17#	

ICC = Intraclasss correlation coefficient; SEM = Standard error of mean; # = data compiled from Alaranta et al. 1994; * = data compiled from Moreland et al. 1997.

TABLE 11.13

Prone Bridging Endurance Mean Values (±SD) in Asymptomatic and Symptomatic Clients

	Mean (s) ± SD
Asymptomatic	
Female	51.2 ± 19.9
Male	92.9 ± 29.3
Fatigue	72.3 ± 33.0
Pain	79.1 ± 33.2
Symptomatic	
Female	24.3 ± 27.5
Male	33.4 ± 26.0
Fatigue	30.6 ± 26.0
Pain	15.4 ± 10.4

Data from Schellenberg et al. 2007.

TABLE 11.14

Supine Bridging Endurance Mean Values (±SD) in Asymptomatic and Symptomatic Clients

	Mean (s) ± SD
Asymptomatic	
Female	152.0 ± 30.2
Male	188.0 ± 45.7
Fatigue	179.3 ± 50.7
Pain	155.6 ± 20.6
Symptomatic	
Female	75.7 ± 44.8
Male	77.9 ± 55.4
Fatigue	73.5 ± 47.0
Pain	80.3 ± 52.3

Data from Schellenberg et al. 2007.

Upper Extremity Testing

It is not uncommon for clients to need to utilize components of muscular strength, muscular power, and endurance of the upper extremities as we have already discussed in relation to the lower extremities. Many athletic skills require generating or transferring explosive power through the upper extremities and trunk musculature (Stockbrugger & Haennel 2001). Certainly activities such as throwing in baseball, the overhand serve in tennis, striking an opponent in boxing, and blocking in football all require strong powerful movements in the upper extremity. Although there are not as many functional tests for the upper extremities as for the lower extremities, in this chapter we describe a number of excellent tests that can be used for clients who require assessment of upper extremities. These assessments include weight-bearing and non–weight-bearing tests as well as tests that involve integration of the upper body, trunk, and lower extremities.

FLEXED ARM HANG

▶ **Purpose:** The flexed arm hang is an endurance test used to assess the forearm and upper arm flexor muscles.

▶ **Equipment:** A pull-up bar high enough that the client's feet cannot touch the ground, stopwatch.

Procedure (Ellenbecker et al. 2000)

1. Use a bar that is slightly higher than the client's standing height.
2. Ask the client to grasp the bar with an overhand grip.
3. Lift the client into place. As the starting position, use a point at which the client's chin is raised above the bar but not touching it; the arms are flexed, and the chest is close to the bar.
4. Release the client and start the stopwatch. The client attempts to keep the chin above the bar as long as possible without excessive body movements.
5. End the test when the chin falls below the level of the bar. Record time to the nearest second.

Analysis and Interpretation of Data

Percentile scores for the flexed arm hang for boys and girls are shown in tables 12.1 and 12.2 on pages 255 and 256.

Statistics

None available.

PULL-UP TEST

▶ **Purpose:** The pull-up test is an endurance test used to assess the forearm and upper arm flexor muscles.

▶ **Equipment:** Standard pull-up bar, stopwatch.

Procedure

1. Ask the client to grasp the bar with an overhand grip.
2. Have the client take the beginning position, hanging with the arms straight (first position).
3. Ask the client to pull the body superior enough that the chin is over the top of the bar (second position).
4. Have the client return to the fully extended hanging position with each repetition.
5. Instruct the client to minimize movements, including swinging of the body.

6. If necessary, you can hold your extended arms in front of the client's thighs to prevent excessive swinging.

Analysis and Interpretation of Data

Score can be given as the total number of pull-ups in a pre-established required amount of time (according to the clinician's preference) or the total number performed until exhaustion.

Statistics

No reliability or validity data are available for this assessment.

First position.

Second position.

ALTERNATE PULL-UP TEST

▶ **Purpose:** The alternate pull-up test is an endurance test used to assess the forearm and upper arm flexor muscles in those who are unable to perform the standard pull-up test.

▶ **Equipment:** Standard pull-up bar, stopwatch.

Procedure (Ellenbecker et al. 2000)

1. Have the client supine on the floor.
2. Place the bar at a height 1 to 2 in. (2.5 to 5 cm) out of reach directly above the client's shoulders.
3. Have an elastic band suspended from the uprights, parallel to the floor, 7 to 8 in. (18 to 20 cm) below the bar.
4. Tell the client that during the pull-up, only the heels should have contact with the floor; the buttocks should be off the floor and the arms and legs should be straight.
5. Ask the client to grasp the bar with an overhand grip and arms fully extended (first position).
6. Count a single repetition when the client can reach the chin over the elastic band (second position).
7. Check that the client performs movements with the arms only; the body should remain rigid and straight.
8. Do not allow any swinging.
9. Have the client perform the pull-up as many times as possible until exhaustion.

Analysis and Interpretation of Data

Score can be given as the total number of pull-ups in a pre-established required amount of time (according to the clinician's preference) or the total number performed until exhaustion.

Statistics

No reliability or validity data are available for this assessment.

First position.

Second position.

PUSH-UP TEST

▶ **Purpose:** To assess upper extremity muscular strength and endurance.

▶ **Equipment:** Stable testing surface.

Procedure (Roetert and Ellenbecker 1994)

1. Have the client begin in the prone position with hands shoulder-width apart and full weight of the lower body on the toes.

2. Instruct the client to begin with the arms extended while keeping the head, shoulders, back, hips, knees, and feet in a straight line.

3. Count a complete push-up when the client's upper arm is parallel to the floor or lower in the down position and the arms are completely extended in the up position and straight body alignment is maintained.

Analysis and Interpretation of Data

The number of push-ups done in 1 min is used as a score and can be ranked according to standards listed in table 12.3 on page 257.

Statistics

No reliability or validity data are available for this assessment.

UNDERKOFFLER SOFTBALL THROW FOR DISTANCE

▶ **Purpose:** To determine muscular power and strength via a functional measurement.

▶ **Equipment:** Large field or area for throwing, softball, tape measure if the field is not marked.

Procedure (Collins and Hedges 1978)

1. Have the client perform warm-up throws until he or she feels able to throw a maximal distance.

2. Following the warm-up throws, have the client perform four more successive warm-up throws, 25%, 50%, 75%, and 100% in intensity.

3. Then give the client three maximal attempts.

4. Measure, to the nearest ½ in. (1.3 cm), the distance of the throw from the client's foot to the location of the ball's first bounce.

Analysis and Interpretation of Data

Score is the distance of throw measured to the nearest location of the ball's first bounce.

Statistics

No reliability or validity data are available for this assessment.

MEDICINE BALL TOSS

▶ **Purpose:** To determine upper extremity strength and power.

▶ **Equipment:** 6 lb (2.7 kg) medicine ball, measuring tape.

Procedure

1. Instruct the client to stand one step behind the starting line.
2. Have the client hold the medicine ball.
3. Ask the client to take one step and toss the ball, staying behind the start line.
4. You can choose the throw that is generally used, which is from chest level, or choose from the following alternatives:
   - A throw simulating the forehand or backhand in tennis.
   - The overhead throw—the client stands behind a line facing forward, holding the medicine ball, and tosses the ball as far as possible from an overhead position with the use of a single step.
   - The reverse throw—the client stands with his or her back to the line and tosses the medicine ball from an underhand position over the shoulder backward.
5. For the score, measure from the front of the line to the point where the ball lands.

Analysis and Interpretation of Data

■ Normative data and correlations to vertical jump height and trunk strength are shown in tables 12.4 through 12.8 on pages 257 through 260.

■ These throws, when used for tennis testing, can be done so as to simulate a forehand and or backhand stroke. Ikeda and colleagues (2007) also describe a side medicine ball throw. When assessing what other factors may account for scores with the side medicine ball throw, Ikeda and colleagues (2007) found that this test may be useful for examining trunk rotation strength in males, but not in females (table 12.4).

■ Aussprung and colleagues (1995) reported that college-age football players who used an 11 lb (5 kg) medicine ball threw an average of 19 to 22 ft (5.8-6.7 m), and those who used the 5 lb (2.3 kg) ball threw an average of 26 to 29 ft (8-8.8 m). These distances were significantly greater than those of female college athletes from various sports, who threw the 5 lb medicine ball an average of 9 to 12 ft (2.7-3.7 m). These throws were done in the seated position on the floor.

■ Salonia and colleagues (2004) assessed the medicine ball throw in 60 female gymnasts and obtained average scores as listed in table 12.8 on page 260.

Statistics

Significant correlations were reported between this test and isometric maximal trunk rotation torque for males, but not for females (table 12.5). Additionally, Ellenbecker and Roetert (2004) found that with elite tennis players, symmetric isokinetic torso rotation strength and isokinetic trunk rotation strength were correlated with a trunk rotation throw using a 2.7 kg (6 lb) medicine ball ($r = 0.787$-0.836, $P < 0.01$) (table 12.6). Ellenbecker and Roetert (2004) have also listed normative values for various groups of tennis players that can be found in table 12.7.

BACKWARD OVERHEAD MEDICINE BALL THROW TEST

▶ **Purpose:** To determine upper extremity strength and power.

▶ **Equipment:** Weighted medicine ball, room to throw medicine ball, tape to mark starting line, measuring tape.

Procedure (Salonia et al. 2004)

1. Ask the client to stand backward, with the feet shoulder-width apart and the heels on the starting line.
2. Have the client hold the medicine ball with the arms straight out in front of the body at shoulder height.
3. You may allow the client to use a countermovement, flexing at the hips and knees and at the same time flexing the trunk forward and lowering the medicine ball to just below waist or hip height.
4. Have the client, after the countermovement, thrust the hips forward and extend the knees and trunk.
5. Have the client flex the shoulders, elevating the ball back up to shoulder height and beyond as it is thrown back over the head.
6. Instruct the client to maintain the arms in an extended position.
7. Monitor the following specific points:
   - During the countermovement, the client should not bend or flex the knees more than 90°.
   - The shoulders should maintain at least 45° of flexion in relation to the trunk.
   - At the end of the throw, the client's feet may leave the ground, as would occur with a jumping motion, to minimize any deceleration component of the vertical ground reaction forces.
   - The client should keep the arms as straight as possible while throwing the ball back over the head with a pendulum action.

Analysis and Interpretation of Data

- Normative values for the overhead medicine ball throw can be seen in tables 12.7 through 12.10 on pages 259 through 261.
- Mayhew and colleagues (2005) determined values for the backward overhead medicine ball throw for

distance in a group of 40 intercollegiate football players following an eight-week off-season conditioning program. The mean distance thrown by the football players was 10.41 m ± 1.45 with a range of 8.17 to 13.85 m. Another important finding in this study was that a large percentage of the football players (55%) had their best throw on the third trial, indicating a substantial learning effect. Interclass correlation

(continued)

(continued from previous page)

coefficient for reliability was 0.86. Because of the learning effect, the authors suggest that there may be a concern with use of this test as an indicator of total-body explosive power if the movement has not been a part of the standard training regime.

■ Stockbrugger and Haennel (2003) examined the factors contributing to the performance of the backward overhead medicine ball throw with both jumping athletes (male volleyball players) and nonjumping athletes (male wrestlers). The absolute values for the overhead medicine ball throw were higher for the jump athletes (15.4 ± 1.1, range 13.4-17.4 m) as compared to the nonjump athletes (14.2 ± 1.8, range 12.2-18.8). Both groups showed strong correlations between the overhead medicine ball throw and the power index ($r = 0.817$ jump, $r = 0.917$ nonjump). The backward overhead medicine ball throw also was highly correlated with the countermovement vertical jump ($r = 0.899$ jump, $r = 0.945$ nonjump).

Statistics

■ Stockbrugger and Haennel (2001) assessed the backward overhead medicine ball throw in 20 competitive volleyball players. They also assessed a countermovement vertical jump that they used to help determine a power index for each athlete. Following testing there was a strong correlation between the distance of the medicine ball throw and the power index for the countermovement vertical jump ($r = 0.906$, $P < 0.01$). The authors also determined the test–retest reliability for the medicine ball throw ($r = 0.996$, $r^2 = 0.993$, SEE = 0.293, $P < 0.01$). These findings suggest that the medicine ball throw test is a valid and reliable test for assessing power for a total-body movement pattern.

■ Duncan and Al-Nakeeb (2005) have reported that five or six trials of the backward overhead medicine ball throw test may be needed before within-subject variation is at an acceptable level.

SIDEARM MEDICINE BALL THROW

▶ **Purpose:** To determine upper extremity strength and power.

▶ **Equipment:** An open space for throwing, a marked distance or a tape measure to measure throws; medicine ball (these come in a variety of sizes and weights).

Procedure

1. Following a brief warm-up, ask the client to grasp the medicine ball with both hands.
2. Ask the client to rotate the trunk in the direction opposite to the throwing direction as a counter-movement.
3. Have the client follow this motion by rotating the trunk in the throwing direction as he or she attempts to throw the medicine ball as far as possible.

4. Record the distance thrown as the score.

Analysis and Interpretation of Data

■ Normative data for the sidearm medicine ball throw for distance are shown in table 12.4 on page 257.

Statistics

Ikeda and colleagues (2007) found that the side medicine ball toss had significant correlations with isometric trunk rotation torque ($r = 0.596\text{-}0.739$, $P < 0.05\text{-}0.01$) and scores for 1-repetition maximum bench press ($r = 0.683\text{-}.0725$, $P < 0.01$).

SEATED CHEST PASS

▶ **Purpose:** To determine upper extremity strength and power.

▶ **Equipment:** 400 g (0.9 lb) netball, measuring tape.

Procedure (Cronin & Owen 2004)

1. Ask the client to assume a seated position on the floor with the lower back, shoulders, and head supported against the wall.

2. Have the client extend the knees.

3. Instruct the client that he or she will hold the netball in both hands and forcefully throw it as far as possible without allowing the head, shoulders, and hips to move from the wall.

4. Have the client perform several warm-up sessions and several practice trials.

5. Give a 1 min rest period.

6. For the test, have the client throw the netball four times; a minimum of 30 to 45 s of rest is recommended between each throw.

7. Score by averaging the distance thrown in the four trials.

Analysis and Interpretation of Data

Score is the average distance of the four throws.

Statistics

No reliability or validity data are available for this assessment.

SEATED SHOT-PUT THROW

▶ **Purpose:** To determine upper extremity strength and power.

▶ **Equipment:** 4.5 kg (10 lb) indoor shot put, adequate room for throwing, measuring tape.

Procedure (Mayhew et al. 1997)

1. Ask the client to assume a seated position on the floor with his or her back against a wall.

2. Have the client flex the knees at approximately a right angle; the feet should apply a backward force to hold the back firmly against the wall.

3. Instruct the client to grasp the shot with both hands and forcefully push it from the center of the chest.

4. Have the client perform several warm-up sessions and several practice trials.

5. For the test, have the client throw the shot three times.

6. Have the client rest between each throw (a minimum of 30 to 45 s of rest is recommended).

7. For scoring, average the distance thrown in the three trials.

Analysis and Interpretation of Data

■ Normative data for performance of the seated shot-put throw and correlations to other measures are listed in tables 12.11 and 12.12 on pages 261 and 262.

Statistics

■ Reliability coefficients of 0.84 have been reported for college-age men (Johnson & Nelson 1979). Johnson and Nelson (1979) also obtained a validity coefficient of 0.77 by correlating distance scores with scores computed by a power formula. Mayhew and colleagues (1995) found intraclass reliability correlation coefficients for the 8 lb (3.6 kg) shot at $r = 0.98$ and for the 10 lb (4.5 kg) shot at $r = 0.95$.

■ Mayhew and colleagues (1991) found that seated shot-put throw was significantly correlated with bench press power 30% ($r = 0.67$) and bench press power 60% ($r = 0.75$). These authors (1993) reported that the seated shot-put throw was significantly

related to absolute power ($r = 0.51$) and relative bench press power output ($r = 0.66$). When they removed the effect of differences in body mass, the effects were reduced dramatically (to absolute power $r = 0.17$ and relative bench press power output $r = 0.29$). They concluded that the seated shot-put test appears to be moderately related to both absolute power and relative bench press power in college football players and may be greatly influenced by size and muscularity. See table 12.11 on page 261 for seated shot-put performance of football players.

■ Mayhew and colleagues (1994) determined the degree of specificity among female athletes regarding the seated shot-put test. When assessing 60 female college athletes, they found that the physical characteristics most highly related to the seated shot-put test were height, body mass, and fat-free mass. They also found high correlations between the Margaria-Kalamen test and the seated shot-put test.

(continued)

(continued from previous page)

■ While assessing strength and power of high school athletes, Mayhew and colleagues (1995) found that both boys and girls threw significantly farther with the 8 lb than with the 10 lb shot. They also deduced that seated shot-put throw scores were significantly related to bench press power tests in boys but not in girls. Additionally, both weights of the seated shot-put test were significantly related to body mass and fat-free mass in boys but not in girls. It may simply be the case that the seated shot-put test relies on a combination of neuromuscular skill, muscle mass, neural activation, and fiber type distribution for success.

Note

Others performing this test have either controlled for the angle of trajectory by limiting the release point or allowed an uncontrolled release point. Angle of release can be controlled by a stand constructed from square tubing from a football goal post. Testing is similar to that described earlier. The client sits with the hips against the back of the chair and strapped into the chair to prevent use of the lower body. The client holds the shot with both hands placed against the center of the chest; the forearms are in the position that is parallel to the floor. The distance from the tip of the elbow to the floor is determined using a tape measure. The height of the crossbar is adjusted to coincide with the height of the elbow. The client tosses the shot forward from the center of the chest as far as possible using both hands. The shot must pass between the post in order to count for a score. The distance from a line extending across the front of the chair to the mark for the shot is measured to the nearest centimeter. One can perform the test with or without controlling the arc of release. Gillespie and Keenum (1987) found that with the angle of release controlled, test and retest reliability coefficients were 0.96 and 0.97. With the angle of release uncontrolled, they obtained reliability coefficients of 0.95 and 0.96.

CLOSED KINETIC CHAIN UPPER EXTREMITY TEST

▶ **Purpose:** The Closed Kinetic Chain Upper Extremity Test (CKCUTEST) is a modification by Davies and Dickoff-Hoffman (1993) of a standard push-up test. This test measures strength, endurance, and closed chain kinetic stability of the upper extremities.

▶ **Equipment:** Stable floor, two pieces of athletic tape, stopwatch.

Procedure (Davies and Dickoff-Hoffman 1993; Goldbeck & Davies 2000; Ellenbecker & Davies 2001)

1. Place the two pieces of athletic tape 3 ft (0.9 m) apart on the floor.
2. Instruct the client to assume the standard push-up position with the hands just inside the two pieces of tape.
3. Instruct the client to move the hands as rapidly as possible from one tape line to the other, touching each line alternately in a "windshield wiper" fashion.
4. Using the stopwatch, count the number of touches in 15 s.
5. You can record scores on the Closed Kinetic Chain Upper Extremity Stability Test (CKCUTEST) form on page 280 in the appendix. (For your convenience, the form is also included on the DVD that accompanies this book.)

Analysis and Interpretation of Data

■ One can normalize data for males and females by dividing the number of line touches by the subject's height in inches.

■ To normalize to height, one can place the two lines a distance apart equal to a specified percentage of the client's height. This may be somewhat inefficient since the distance would change with each person to be tested.

■ Rousch and colleagues (2008) assessed a sample of 77 uninjured collegiate male baseball players to determine reference values for the CKCUTEST. They found no difference in scores according to player position. For a summary of normative values, see table 12.13 on page 262.

■ One can determine power scores by multiplying 68% of the subject's body weight (the superincumbent weight of the upper extremities, head, and trunk) in pounds by the number of touches and dividing by 15 s:

Power = (68% body weight (lb) × Touches) / 15 s.

Statistics

Test–retest reliability was found to be 0.927 (Goldbeck & Davies 2000).

FUNCTIONAL THROWING PERFORMANCE INDEX (FTPI)

▶ **Purpose:** To determine accuracy of throwing. This index was originally described by Davies and Dickoff-Hoffman (1993).

▶ **Equipment:** Tape, a wall, a 21 in. (53 cm) rubber playground ball.

Procedure

1. Place a piece of tape 15 ft (4.6 m) from a wall. On the wall, 4 ft (1.2 m) above the floor, place a 1 ft by 1 ft (0.3 m by 0.3 m) square.

2. Give the client four submaximal- to maximal-gradient warm-up throws.

3. Then give the client 30 s to perform as many accurate throws as possible into the square.

Analysis and Interpretation of Data

In most cases three sets are performed, and the score is the percentage accuracy: the total number of throws / the number of accurate throws × 100.

Statistics

Reliability shows intraclass correlation coefficients (ICCs) above 0.90 (Davies & Dickoff-Hoffman 1993). Norms as described by Davies & Dickoff-Hoffman (1993) are as follows: Males = 15 throws, 7 accuracy, for a FTPI at 47% (range 33-60%); Females = 13 throws, 4 accuracy, for a FTPI of 29% (range 17-41%).

Note

Specific throwing parameters are often difficult to assess, especially without video analysis. A trained eye is necessary to closely monitor for specific dysfunction. The Upper Extremity Throwing Analysis form on page 281 in the appendix gives the clinician some guidelines to use when assessing a client. (For your convenience, the analysis form is also included on the DVD that accompanies this book.) It is advisable to monitor multiple throws from various positions and if possible to use video analysis to look critically at the client's technique.

TABLE 12.1

Flexed Arm Hang for Boys

Percentile	AGE											
	6	7	8	9	10	11	12	13	14	15	16	17+
100	55	95	63	101	120	101	111	127	117	130	125	116
95	23	60	34	40	48	52	47	48	68	79	71	64
90	16	23	28	28	38	37	36	37	61	62	61	56
85	14	20	23	24	31	31	30	33	47	58	51	49
80	12	17	18	20	25	26	25	29	40	49	46	45
75	10	15	17	18	22	22	21	25	35	44	42	41
70	9	13	15	16	20	19	19	22	31	40	39	39
65	9	11	14	14	17	17	16	20	28	37	36	37
60	8	10	12	12	15	15	15	18	25	35	33	35
55	7	9	11	11	14	13	13	16	22	33	30	33
50	6	8	10	10	12	11	12	14	20	30	28	30
45	5	7	9	8	10	10	10	12	17	28	25	29
40	5	6	8	8	8	9	9	10	15	25	22	26
35	4	5	6	7	7	7	8	9	13	22	20	23
30	3	4	5	5	6	6	6	8	11	20	18	20
25	2	4	4	5	5	5	5	6	10	18	15	17
20	2	3	3	3	3	4	4	5	8	14	12	15
15	1	2	2	3	2	3	2	4	5	10	10	11
10	1	1	1	2	1	1	1	2	3	8	7	8
5	0	0	0	0	0	0	0	0	1	3	3	5
0	0	0	0	0	0	0	0	0	0	0	0	0

Percentile scores are based on age and test scores in seconds.

The President's Challenge Physical Activity and Fitness Awards Program, a program of the President's Council on Physical Fitness and Sports, U.S. Department of Health and Human Services.

TABLE 12.2

Flexed Arm Hang for Girls

Percentile	AGE											
	6	7	8	9	10	11	12	13	14	15	16	17+
100	55	72	97	78	152	150	99	68	100	125	131	127
95	22	29	26	35	38	33	37	35	38	41	40	37
90	15	21	21	23	29	25	27	28	31	34	30	29
85	13	17	17	20	22	20	21	21	25	28	24	24
80	11	14	15	16	19	16	16	19	21	23	21	20
75	10	12	13	14	16	14	14	16	18	18	18	18
70	9	11	11	12	14	13	13	14	16	15	16	15
65	8	9	10	11	12	11	11	12	13	12	13	12
60	6	8	10	10	11	9	10	10	11	10	10	11
55	6	7	9	9	9	8	8	9	10	9	9	10
50	5	6	8	8	8	7	7	8	9	7	7	7
45	5	5	7	7	7	6	6	6	7	6	6	6
40	4	5	6	6	6	5	5	5	6	5	5	5
35	3	5	5	5	5	4	4	5	5	4	4	5
30	3	4	4	4	4	4	3	4	4	4	3	4
25	2	3	3	3	3	3	2	3	3	3	2	2
20	1	2	3	2	2	2	1	1	2	2	2	2
15	1	1	1	1	1	1	1	1	1	1	1	1

Percentile scores are based on age and test scores in seconds.

The President's Challenge Physical Activity and Fitness Awards Program, a program of the President's Council on Physical Fitness and Sports, U.S. Department of Health and Human Services.

TABLE 12.3

Push-Up
(Number Completed in 1 min)

	Excellent	Good	Average	Needs improvement
FEMALE				
Adult	>44	34-44	27-36	<24
Junior	>42	34-42	20-34	<20
MALE				
Adult	>49	40-49	30-40	<30
Junior	>52	49-52	35-49	<35

Adult is defined as >18 years of age and junior as up to 18 years of age.

Reprinted, by permission, from P. Roetert and T. Ellenbecker, 1998, *Complete conditioning for tennis* (Champaign, IL: Human Kinetics), 16.

TABLE 12.4

Side Medicine Ball
Throw Distance (m)

Weight and side	Males (*n* = 16), mean ± SD	Females (*n* = 10), mean ± SD	% difference, mean ± SD
2 kg right	15.4 ± 2.1	11.0 ± 1.0	29
2 kg left	15.2 ± 2.2	10.7 ± 0.9	30
4 kg right	11.0 ± 1.8	8.0 ± 0.7	27
4 kg left	11.2 ± 2.0	8.0 ± 0.7	29
6 kg right	9.0 ± 1.4	6.4 ± 0.5	29
6 kg left	8.9 ± 1.7	6.2 ± 0.4	30

Data from Y. Ikeda et al., 2007, "Relationship between side medicine-ball throw performance and physical ability for male and female athletes," *European Journal of Applied Physiology* 99:47-55.

TABLE 12.5

Correlations of Side Medicine Ball Throw
to Selected Variables Males and Females

Variable	MALES (*n* = 16) 2 kg	4 kg	6 kg	FEMALES (*n* = 10) 2 kg	4 kg	6 kg
Height	0.648	0.651	0.697	0.581	0.187	0.430
Body mass	0.289	0.274	0.358	0.796	0.489	0.697
Isometric maximum trunk torque (R)	0.470	0.596	0.739	0.196	−0.001	0.072
Isometric maximum trunk torque (L)	0.611	0.636	0.661	0.478	0.429	0.239
1RM bench press	0.586	0.658	0.682	0.774	0.855	0.848
I RM parallel squat	0.683	0.717	0.725	0.279	0.324	0.476
Bench press peak power	0.642	0.693	0.729	0.860	0.723	0.845
Static squat jump peak power	0.553	0.461	0.496	0.394	0.236	0.605
Vertical jump	0.161	0.246	0.207	0.249	0.349	0.329

Adapted from Y. Ikeda et al. 2007. "Relationship between side medicine-ball throw performance and physical ability for male and female athletes." *European Journal of Applied Physiology*;99:47-55.

TABLE 12.6

Pearson Correlations Between Isokinetic Trunk Rotation Strength and Functional Medicine Ball Toss

Parameter	Forehand	Backhand
FH PT 60	0.836	0.814
FH PT 120	0.830	0.820
FH W 60	0.815	0.785
FH W 120	0.815	0.797
FH TW 120	0.831	0.813
FH Endurance	0.877	0.788
BH PT 60	0.827	0.787
BH PT 120	0.825	0.792
BH W 60	0.820	0.779
BH W 120	0.834	0.798
BH TW 120	0.855	0.813
BH Endurance	0.018 NS	0.030 NS

All correlations are significant at the 0.01 level except those noted as NS (nonsignificant). FH = forehand (left rotation); BH = backhand (right rotation); PT = peak torque; W = work; TW = total work.

Reprinted, by permission, from T.S. Ellenbecker and E.P. Roetert, 2004, "An isokinetic profile of trunk rotation strength in elite tennis players," *Medicine and Science in Sports and Exercise*, 36:1959-1963.

TABLE 12.7

Medicine Ball Throw Scores

	Excellent	Good	Average	Needs improvement
Forehand (ft)				
Female, adult	>30.5	25-30.5	19.5-25	<19.5
Female, junior	>32	26-32	20-26	<20
Male adult	>39	32-39	25-32	<25
Male junior	>42	35-42	28-35	<28
Backhand (ft)				
Female adult	>30	24-30	17.5-23.5	<17.5
Female junior	>31	25-31	18-25	<18
Male adult	>37.5	30.5-37.5	23.5-30.5	<23.5
Male junior	>42	34-42	26-34	<26
Overhead (ft)				
Female adult	>22.5	18.5-22.5	14.5-18.5	<14.5
Female junior	>23	19-23	15-19	<15
Male adult	>30.5	25.5-30.5	20-26.5	<20
Male junior	>34	29-34	23-29	<23
Reverse Overhead (ft)				
Female adult	>32.5	26.5-32.5	20.5-26.5	<20.5
Female junior	>34	27-34	20-27	<20
Male adult	>43.5	35-43.5	27-35	<27
Male junior	>46	38-46	31-38	<31

Adult is defined as >18 years of age and junior as up to 18 years of age.

Reprinted, by permission, from T.S. Ellenbecker and E.P. Roetert, 2004, "An isokinetic profile of trunk rotation strength in elite tennis players," *Medicine and Science in Sports and Exercise,* 36:1959-1963.

TABLE 12.8

Throw Means Categorized by Age, Class Level, and Throw Type

	LEVELS 5 AND 6			LEVELS 7 AND 8		
	OF	OB	CP	OF	OB	CP
Age 10 years	164.75	135.13	131.00	126.25	108.13	124.13
	168.25	166.75	107.75	146.13	105.25	145.25
	112.25	119.13	150.75	139.25	135.38	138.25
	150.63	107.25	149.38	159.38	128.25	128.88
	141.63	151.88	137.88	147.75	192.13	140.75
Age 11 years	153.63	185.38	122.13	160.25	136.38	125.38
	127.75	145.63	131.13	148.75	159.38	159.63
	179.50	179.38	125.13	237.25	137.75	168.88
	133.50	160.13	149.88	152.13	125.63	128.88
	135.50	137.00	122.13	158.13	139.75	147.75

OF= overhead forward throw; OB = overhead backward throw; CP = chest pass. Levels describe gymnast's ability and rank. Level 1 is the lowest possible level; level 5 is the point at which gymnasts begin to compete. State, national, and Olympic-ranked gymnasts are at level 10.

Reprinted, by permission, from M.A. Salonia et al., 2004, "Upper body power as measured by medicine ball throw distance and its relationship to class level among 10- and 11-year-old female participants in club gymnastics," *Journal of Strength and Conditioning Research* 18(4):695-702.

TABLE 12.9

Normative Values for Elite Weightlifters for the Overhead Medicine Ball Throw (m)

	BODY WEIGHT (KG)									
Test	52	56	60	67.5	75	82.5	90	100	110	+110
Backward overhead weight throw (7.5 kg; juniors 5.0 kg)	11	12	13	14.5	14.5	14.5	14.5	15	15	14.5

Adult is defined as >18 years of age and junior as up to 18 years of age.

Data from Ajan and Baroga, 1988.

TABLE 12.10

Normative Data for Backward Overhead Medicine Ball Throw
(NCAA Division I and II Athletes, $n = 42$)

Test	FEMALES			MALES		
	Average	Good	Elite	Average	Good	Elite
Backward overhead medicine weight throw	6.7-9.1 m	9.2-12.2 m	12.2-15.2 m	11.6-12.8 m	12.8-16.4 m	16.4-19.8 m

Data from Field, 1989 and Field, 1991.

TABLE 12.11

Physical Performance Characteristics for the Seated Shot-Put Test

Study	Mean	SD	Range
Mayhew et al. 1991 Weightlifting class ($n = 46$)	4.09 m	0.63 m	17.0 m to 24.0 m
Mayhew et al. 1993 College football players ($n = 40$)	4.7 m	0.5 m	3.7 m to 5.9 m
Mayhew et al. 1994 College female athletes ($n = 64$)	2.88 m	0.34 m	2.27 m to 3.61 m
Mayhew et al. 1995 High school athletes Male ($n = 36$) 8 lb shot 10 lb shot Female ($n = 23$) 8 lb shot 10 lb shot	4.07 m 3.72 m 2.98 m 2.64 m	0.81 m 0.68 m 0.32 m 0.32 m	Not reported Not reported Not reported Not reported

Based on Y. Ikeda et al., 2007, "Relationship between side medicine-ball throw performance and physical ability for male and female athletes," *European Journal of Applied Physiology;* 99:47-55.

TABLE 12.12

Correlations to Seated Shot-Put Test

STUDY AND PARAMETER		CORRELATION (SD)		
	Mayhew et al. 1991	Mayhew et al. 1994 (8 lb)	Mayhew et al. 1995 (boys 8 lb, 10 lb; girls 8 lb, 10 lb)	
PHYSICAL PARAMETERS				
Height (cm)		0.48	0.62, 0.72	0.33, 0.45
Body mass (kg)		0.61	0.670, 0.73	0.29, 0.49
Fat-free mass (kg)		0.66	0.24, 0.29	0.34, 0.49
% fat		0.18	0.24, 0.29	0.05, 0.36
OTHER POWER TESTS				
Margaria-Kalamen test		0.65 (0.37)		
Vertical jump		0.26 (0.49)		
Bench press power		0.38 (0.41)		
Bench press (kg)	0.73			
Bench press power (30%)	0.67			
Bench press power (60%)	0.75			

TABLE 12.13

Descriptive and Normative Data for CKCUTEST

Study	Subjects	Males (SD)	Females (SD)
Goldbeck & Davies 2000	24	27.8 (1.77)	ND
Ellenbecker et al. 2000	ND	18.5 (ND)	20.5 (ND)
Roush et al. 2008	77	30.41(3.87)	ND

ND = not described.

Lower Extremity Anaerobic Power Testing

Power has previously been explained in this text as force times distance or work divided by the time needed to perform the activity. The energy system most often recruited in truly powerful movements (vertical jump, horizontal jump, and other tests listed in chapter 9) is the adenosine triphosphate–phosphocreatine (ATP-PC) system. The tests listed in this chapter are often referred to as power tests, but they typically involve the anaerobic energy system. Anaerobic power is the ability to perform high-intensity exercise for durations lasting from a fraction of a second to several minutes (Hoffman 2006). Therefore, these tests typically last longer than many of the tests listed in chapter 9, although some tests in both this chapter and chapter 9 could be discussed in either chapter. We again encourage clinicians to look critically at the client's requirements for successful performance of the given task and determine the most appropriate test(s).

300 SHUTTLE RUN

▶ **Purpose:** Assessment of anaerobic lower extremity power performed repetitively over a fixed-distance course.

▶ **Equipment:** Stopwatch, flat nonslip 25 yd (23 m) course marked with tape or cones at each end (two courses if two individuals will be tested at one time).

Procedure (Gilliam 1983; Harman et al. 2000)

1. Position the client directly behind the starting line.
2. Have the client sprint to the 25 yd line to touch the line with the foot, then turn to sprint back to the starting line and touch that line for a total of six round trips (i.e., 12 × 25 yd = 300 yd [274 m]).
3. Record the time to the nearest tenth of a second.
4. Have the client rest 5 min and then repeat the test.
5. Record the average of the two tests as the client's time.

Analysis and Interpretation of Data

■ The two tests should be close to each other in time in order to demonstrate that the client has good recovery and speed endurance. If the separate test times are not in close proximity, their reliability should be questioned (Gilliam 1983).

■ Percentile ranks of Division I athletes for this test are listed in table 13.1 on page 271.

Statistics

There are no reliability or validity data available for this assessment.

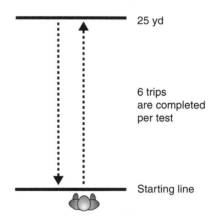

25 yd

6 trips are completed per test

Starting line

RUNNING-BASED ANAEROBIC SPRINT TEST (RAST)

▶ **Purpose:** Assessment of anaerobic lower extremity sprint power performed repetitively over a fixed-distance course.

▶ **Equipment:** Stopwatch, flat nonslip 35 m course marked with tape or cones at each end (two courses if two individuals are to be tested at one time).

Procedure

1. Have the client positioned directly behind the starting line.
2. For the test, require the client to perform six 35 m discontinuous maximal-effort sprints.
3. Allow 10 s for rest at the turnaround between each sprint.
4. Start the stopwatch as the client begins.
5. Each time the client completes the 35 m distance, stop the stopwatch for the 10 s rest and then start it again.

Analysis and Interpretation of Data

■ Power output: Weight of client (kg) × Distance (m^2) / Time (s^3).

■ Fatigue index: (Maximum power − Minimum power) / Total time for six sprints (s).

Statistics

Test–retest reliability for this test: $r = 0.90$ (Balciunas et al. 2006).

LOWER EXTREMITY FUNCTIONAL TEST

▶ **Purpose:** The Lower Extremity Functional Test (LEFT) is a comprehensive timed test of eight multidirectional skills performed continuously in a standardized 16-step sequence between targets. The test incorporates the skills of forward and retro running, side shuffling, carioca, figure 8 running, 45° and 90° cutting, and 90° crossover cutting.

▶ **Equipment:** An area large enough to contain a 30 by 10 ft course (9 by 3 m) configured into the shape of a diamond; four standardized 1 ft (0.3 m) triangular targets fabricated from athletic tape to mark the boundaries of the course; standard stopwatch.

Procedure (Davies & Zillmer 2000)

1. Have the client positioned directly behind the starting line (point A).

2. Ask the client to run through the course following the 16 steps shown in the diagram.

3. Although this test is timed, you should also watch the client for signs of discomfort or compensation.

4. Record the time to the nearest second.

Analysis and Interpretation of Data

▪ The LEFT is a timed test in which improved function is seen with faster times.

▪ Normative values for this test are listed in table 13.2.

Statistics

Tabor and colleagues (2002) established intraclass correlation coefficients of 0.95 and 0.97 in two different university settings.

1. Forward sprint (A-C-A)
2. Retro sprint (A-C-A)
3. Side shuffle right – face in (A-D-C-B-A)
4. Side shuffle left – face in (A-B-C-D-A)
5. Cariocas right – face in (A-D-C-B-A)
6. Cariocas left – face in (A-B-C-D-A)
7. Figure 8s right (A-D-C-B-A)
8. Figure 8s left (A-B-C-D-A)
9. 45° Cuts right – plant outside foot (A-D-C-B-A)
10. 45° Cuts left – plant outside foot (A-B-C-D-A)
11. 90° Cuts right – plant outisde foot (A-D-B-A)
12. 90° Cuts left – plant outside foot (A-B-D-A)
13. Crossover 90° cuts right – plant inside foot (A-D-B-A)
14. Crossover 90° cuts left – plant inside foot (A-B-D-A)
15. Forward sprint (A-C-A)
16. Retro sprint (A-C-A)

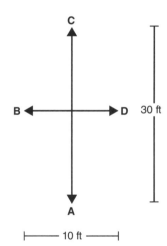

WINGATE ANAEROBIC POWER TEST

▶ **Purpose:** To assess a client's peak power, mean power, and fatigue index.

▶ **Equipment:** Cycle ergometer.

Procedure (Bar-Or et al. 1997; Bar-Or 1987)

1. Have the client sit on the ergometer, and adjust the ergometer to ensure an optimal riding position.
2. Ask the client to warm up for approximately 5 min by performing two to four 5 s sprints.
3. Control the flywheel tension so that tension is applied at the beginning of the test.
4. Encourage the client to pedal as fast as possible prior to the application of resistance.
5. Apply resistance (0.075 kg per kilogram body mass for tests performed on a Monark cycle ergometer).
6. Following the application of the resistance, have the client attempt to pedal at maximum speed for 30 s. Begin the 30 s count only after applying the resistance.

Analysis and Interpretation of Data

Descriptive data and percentile ranks of various athletic populations for this test are listed in tables 13.3 and 13.4 on page 272.

Statistics

Test–retest reliability for the Wingate test varies from 0.89 to 0.99 (Inbar et al. 1996).

BOSCO TEST

▶ **Purpose:** To assess anaerobic lower extremity power via repetitive jumping.

▶ **Equipment:** Large, square 1-dimensional force platform.

Procedure

1. Instruct the client to perform continuous rapid maximal-effort jumps for a duration of 60 s.

2. Instruct the client to lower him- or herself to approximately 90° of knee flexion during the transition from one jump to the next.

3. Ask the client to keep hands on hips throughout the entire session to minimize contributions of the upper body to test performance (Sands et al. 2004).

4. Ask the client to attempt to land in the same position as the starting position to minimize horizontal or lateral displacements.

Analysis and Interpretation of Data

■ This test uses a digital timer located in the testing mat that measures actual flight time. The timer is triggered by the client's feet at the moment of release from the platform and is stopped when the client lands.

■ Additionally, the clinician counts the number of jumps the client can perform during the given testing time. From these data, calculations for mechanical power can be derived.

■ Average and peak power values for this test are listed in table 13.5.

Statistics

The reliability of the Bosco test has been found to be from 0.87 to 0.97 (Bosco et al. 1983; Sands 2000).

SQUAT JUMP TEST

> **Purpose:** To assess lower extremity anaerobic power with a jumping task.

> **Equipment:** Yardstick (or Vertec if available), ladder (if Vertec not available), wall with high ceiling and a good landing area (flat nonslip surface), chalk.

Procedure (Young 1995; Maulder & Cronin 2005)

1. Have the client stand on the foot of the test extremity.

2. Ask the client to sink to hold a knee position of approximately 120° (first position).

3. Instruct the client that at the count of 4, he or she should jump as high as possible either to reach to the Vertec or to mark a spot with the chalk on the wall (second position).

4. Count as successful a jump that is not preceded by sinking or a countermovement.

Analysis and Interpretation of Data

Maulder and Cronin (2005) used a contact mat to measure actual flight times during the squat jump test. They assessed characteristics in 18 healthy male subjects.

Statistics

Normative and reliability values for this test are given in tables 13.6 through 13.8 on page 274.

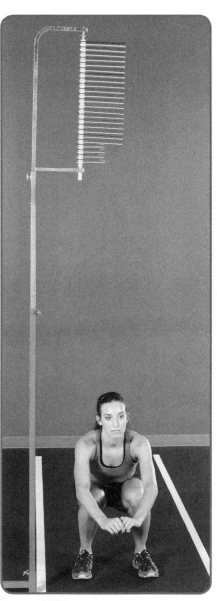

First position.

Second position.

PLYOMETRIC LEAP TEST

▶ **Purpose:** To assess lower extremity anaerobic power via a plyometric type of leaping motion.

▶ **Equipment:** Large nonslip surface and plastic tape measure for recording distance leaped.

Procedure

1. After a brief warm-up, have the client perform three consecutive leaps from a standing position by springing from one foot to the opposite foot.
2. Ask the client to land on both feet after the last leap.
3. Use caution by stressing that the client should land with a bent knee to reduce risk of injury.
4. Record the total distance leaped to the nearest centimeter or meter.

Analysis and Interpretation of Data

No established standards were found.

Statistics

Sinnett and colleagues (2001) studied relationships between field tests of anaerobic power, including the plyometric leap, and 10 km run performance. During their testing of 36 trained runners (20 men, 16 women), they determined that their total subject group on average was able to leap 5.5 m (SD = 0.72). The men leaped 5.9 m (SD = 0.57), and the women leaped 4.9 m (SD = 0.31). For the entire group, there were significant correlations between the plyometric leap and 10 km run times ($r = -0.86$). For both men and women, the plyometric leap was significantly correlated to 10 km run times ($r = -0.778$ and $r = -0.725$, respectively). The study further demonstrated, using multiple regression, that the plyometric leap and 300 m sprint times are the best predictors of 10 km run performance, explaining 78% of the variance in 10 km run times.

Note

Be aware that this test is almost identical to the triple jump in track and field; the difference is that this test starts from a standing position instead of a running start.

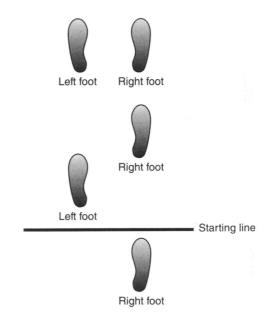

300-METER SPRINT

▶ **Purpose:** To assess lower extremity anaerobic power via sprinting.

▶ **Equipment:** Indoor or outdoor track and standard stopwatch.

Procedure

1. Position the client so that he or she is standing directly behind the starting line.

2. Ask the client to run 300 m at maximal effort.

3. Use the stopwatch to determine the total elapsed time, and record the time to the nearest 1/100 s.

Analysis and Interpretation of Data

No established standards were found.

Statistics:

Sinnett and colleagues (2001) studied relationships between field tests of anaerobic power, including the 300 m sprint, and 10 km run performance. During their testing of 36 trained runners (20 men, 16 women) they found that their total subject group on average was able run 300 m in 56.5 s (SD = 0.6.96). The men ran 53.6 s (SD = 7.12), and the women ran 60.1 s (SD = 4.83). The entire group showed significant correlations between the 300 m run times and the 10 km run times ($r = 0.79$). For both men and women, the 300 m run times were significantly correlated to 10 km run times ($r = 0.719$ and $r = 0.731$, respectively). The study further demonstrated, using multiple regression, that the plyometric leap and 300 m sprint times are the best predictors of 10 km run performance, explaining 78% of the variance in 10 km run times.

TABLE 13.1

Shuttle Run Percentile Ranks (s) in NCAA Division I Athletes

	PERCENTILE RANK (%)									MEAN	*n*
	10	20	30	40	50	60	70	80	90		
Women's basketball	68.9	68.1	66.8	65.9	65.2	64.7	63.6	61.8	58.4	64.7	82
Men's basketball	60.2	58.9	58.1	57.2	56.7	56.3	55.6	55.1	54.1	57.1	125
Softball	78.0	74.6	72.4	71.3	69.2	67.9	66.5	65.1	63.3	70.0	114
Baseball	67.7	65.3	63.9	63.2	62.0	61.3	59.9	58.9	56.7	62.2	107

Adapted, by permission, from G. McKenzie Gilliam, 1983, "300 yard shuttle run," *NSCA Journal* 5:46.

TABLE 13.2

Lower Extremity Functional Test (LEFT)

NORMS		
Males	Females <25 years	Females >25 years
90 s—good	100 s—good	120 s—good
100 s—average	120 s—average	150 s—average
125 s—below average	140 s—below average	180 s—below average

Davies and Zillmer, 2000.

TABLE 13.3

Percentile Ranks for the Wingate Anaerobic Power Test in Physically Active College-Aged Males and Females

	MALES		FEMALE	
Percentile	(W)	(W · kg⁻¹)	(W)	(W · kg⁻¹)
90	662	8.2	470	7.3
80	618	8.0	419	7.0
70	600	7.9	410	6.8
60	577	7.6	391	6.6
50	565	7.4	381	6.4
40	548	7.1	367	6.1
30	530	7.0	353	6.0
20	496	6.6	337	5.7
10	471	6.0	306	5.3

Wingate test performed on a Monark cycle ergometer using a resistance of 0.075 kg · kg body mass⁻¹. Males and females were 18 to 28 years. For males n = 60; for females, n = 69.

Adapted with permission from *Research Quarterly for Exercise and Sport,* Vol. 60. No. 2. 144-151. Copyright 1989 by the American Alliance for Health, Physical Education, Recreation and Dance, 1900 Association Drive, Reston, VA 20191.

TABLE 13.4

Descriptive Data for the Wingate Anaerobic Power Test in Athletic Populations

Population	Gender	PEAK POWER		MEAN POWER		Deviance and resistance	Source
		(W)	(W · kg⁻¹)	(W)	(W · kg⁻¹)		
BASKETBALL							
NCAA DI	F	663 ± 98	9.5 ± 1.4	498 ± 51	7.2 ± 0.7	Bodyguard	LaMonte et al. 1999
Guards		629 ± 79	10.2 ± 1.2	477 ± 45	7.7 ± 0.7		
Forwards		693 ± 106	9.4 ± 1.3	516 ± 49	7.1 ± 0.8		
Centers		668 ± 78	8.3 ± 1.2	502 ± 58	6.3 ± 0.7		
Israel national	M		14.4 ± 1.7		9.1 ± 1.2	Fleish (0.052 kg · kg body mass)	Hoffman et al. 1999
BOBSLED							
U.S. national	M	1005 ± 90	10.8 ± 0.5	796 ± 60	8.6 ± 0.9	Bodyguard	Osbeck, Maiorca, & Rundell 1996

Population	Gender	PEAK POWER (W)	PEAK POWER (W · kg⁻¹)	MEAN POWER (W)	MEAN POWER (W · kg⁻¹)	Deviance and resistance	Source
FOOTBALL							
NCAA DI Backs Linebackers Linemen	M	1189 ± 130 1130 ± 126 1298 ± 83 1223 ± 123	12.2 ± 1.7 13.2 ± 1.1 12.6 ± 0.9 10.4 ± 1.5	874 ± 102 836 ± 91 928 ± 102 879 ± 103	9.1 ± 1.5 10.0 ± 0.9 9.4 ± 1.1 7.5 ± 1.3	Monark (0.083 kg · kg body mass)	Seiler et al. 1990
NCAA DIII	M	1894 ± 140		1296 ± 66		Lode Excalibur (1.1 Nm · kg body mass)	Hoffman et al. 2004
HOCKEY							
U.S. hockey	F	785		729			Unpublished data
KICKBOXING							
Elite	M	1360	18.8	761	10.5	Monark	Zabukovec & Tiidus 1995
MIDDLE DISTANCE RUNNING							
Competitive French	M	842 ± 123	13 ± 2.0	578 ± 64	9.0 ± 1.0	Monark	Granier et al. 1995
SOCCER							
Youth U14 U15 U16	M		9.3 ± 0.2 10.0 ± 0.3 10.5 ± 0.2		8.0 ± 0.2 8.1 ± 0.2 8.7 ± 0.2	Monark (0.075 kg · kg body mass)	Vanderford et al. 2004
U.S. national	M				8.1 ± 0.9	Monark (0.075 kg · kg body mass)	Mangine et al. 1990
SOFTBALL							
Masters	F	406 ± 56					Terbizan et al. 1996
SPEED SKATING							
Elite Canadians	M F		12.3 ± 0.5 16.6 ± 0.9		9.7 ± 0.2 12.7 ± 0.5	Monark (0.092 + 0.112 kg · kg body mass)	Smith & Roberts 1991
SPRINTING (COMPETITIVE)							
Belgians	M	1021 ± 139	14.2 ± 1.4			Monark	Crielaard & Pirnay 1981
French	M	924 ± 105	14 ± 1.0	662 ± 61	10 ± 1.0	Monark	Granier et al. 1995
TENNIS							
NCAA	F	699 ± 130				Monark (0.075 kg · kg body mass)	Kraemer et al. 2003

TABLE 13.5

Average and Peak Power Values for the Bosco Test

	Women average	Men average	Women peak	Men peak
W/kg	12.19	17.81	19.21	23.65
SD	2.39	2.73	6.87	3.08
Watts	922.53	1383.99	1531.76	1845.71
SD	339.42	172.45	10001.73	249.09
Allometric	50.23	74.76	80.80	99.42
SD	11.06	9.58	35.91	11.3

Data from Sands et al. 2004.

TABLE 13.6

Statistical Analyses for the Jump Assessment (10 Subjects)

Variable		Mean (SD)	CV	ICC
Height in meters	Dominant leg	0.157 (0.040)	3.3	0.86 (0.000)
	Nondominant leg	0.158 (0.033)	4.4	0.82 (0.001)

CV = coefficients of variation; ICC = intraclass correlation coefficient
Data from Maulder and Cronin 2005.

TABLE 13.7

Mean (SD) and Symmetry Index for the Squat Jump Test

Dominant	Nondominant	Symmetry index	P-value
1.596 (0.139)	1.617 (0.136)	101 (4)	0.211

Data from Maulder and Cronin 2005.

TABLE 13.8

Intraclass Correlation Coefficients Between Squat Jump and Other Tests

Tests	ICC
Horizontal countermovement jump	0.71
Horizontal squat jump	0.66
Horizontal repetitive jump	0.76
Vertical countermovement jump	0.90

Data from Maulder and Cronin 2005.

appendix

Reproducible Forms

The forms in this appendix will be useful as you administer specific tests to your clients. You may photocopy the forms as needed. For your convenience, the forms are also included on the DVD that accompanies this book.

- Functional Movement Screen Scoring Sheet, page 276
- Tinetti Assessment Tool for Balance, page 277
- Tinetti Assessment Tool for Gait, page 278
- Chester Step Test Graphical Datasheet and Normative Data, page 279
- Closed Kinetic Chain Upper Extremity Stability Test, page 280
- Upper Extremity Throwing Analysis, page 281-283

The Functional Movement Screen Scoring Sheet

Name: _____ Date: _____

Test	Raw score	Final	Comments
Deep squat			
Hurdle step L			
Hurdle step R			
In-line lunge L			
In-line lunge R			
Shoulder mobility L			
Shoulder mobility R			
Active impingement L			
Active impingement R			
Active straight leg raise L			
Active straight leg raise R			
Trunk stability push-up			
Extension screen			
Rotational stability L			
Rotational stability R			
Flexion screen			
Total			

Tinetti Assessment Tool for Balance

Client's name: _____ Date: _____

Location: _____ Rater: _____

Initial instructions: Client is seated in a hard, armless chair. The following maneuvers are tested.

Task	Description of balance	Possible	Score
1. Sitting balance	Leans or slides in chair Steady, safe	= 0 = 1	
2. Arises	Unable without help Able, uses arms to help Able without using arms	= 0 = 1 = 2	
3. Attempts to arise	Unable without help Able, requires >1 attempt Able to rise, 1 attempt	= 0 = 1 = 2	
4. Immediate standing balance (first 5 seconds)	Unsteady (swaggers, moves feet, trunk sway) Steady but uses walker or other support Steady without walker or other support	= 0 = 1 = 2	
5. Standing balance	Unsteady Steady but wide stance (medial heels >4 in. apart) and uses cane or other support Narrow stance without support	= 0 = 1 = 2	
6. Nudged (subject at max position with feet as close together as possible, examiner pushes lightly on subject's sternum with palm of hand 3 times)	Begins to fall Staggers, grabs, catches self Steady	= 0 = 1 = 2	
7. Eyes closed (at maximum position #6)	Unsteady Steady	= 0 = 1	
8. Turning 360°	Discontinuous steps Continuous steps Unsteady (grabs, swaggers) Steady	= 0 = 1 = 0 = 1	
9. Sitting down	Unsafe (misjudges distance, falls into chair) Uses arms or not smooth motions Safe, smooth motion	= 0 = 1 = 2	
		Balance score	

Tinetti Assessment Tool for Gait

Client's name: _____ Date: _____

Location: _____ Rater: _____

Initial instructions: Subject stands with examiner, walks down hallway or across the room, first at usual pace, then back at rapid but safe pace (using usual walking aids).

Task	Description of gait	Possible	Score
10. Initiation of gait (immediately after told to "go")	Any hesitancy or multiple attemps to start	= 0	
	No hesitancy	= 1	
11. Step length and height	a. Right swing foot does not pass left stance foot with step	= 0	
	b. Right foot passes left stance foot	= 1	
	c. Right foot does not clear floor completely with step	= 0	
	d. Right foot completely clears floor	= 1	
	e. Left swing foot does not pass right stance foot with step	= 0	
	f. Left foot passes right stance foot	= 1	
	g. Left foot does not clear floor completely with step	= 0	
	h. Left foot completely clears floor	= 1	
12. Step symmetry	Right and left step length not equal (estimate)	= 0	
	Right and left step appear equal	= 1	
13. Step continuity	Stopping or discontinuity between steps	= 0	
	Steps appear continuous	= 1	
14. Path (estimated in relation to floor tiles, 12 in. diameter; observe excursion of 1 ft over about 10 ft of the course)	Marked deviation	= 0	
	Mild/moderate deviation or uses walking aid	= 1	
	Straight without walking aid	= 2	
15. Trunk	Marked sway or uses walking aid	= 0	
	No sway but flexion of knees or back, or spreads arms out while walking	= 1	
	No sway, no flexion, no use of arms, and no use of walking aid	= 2	
16. Walking stance	Heels apart	= 0	
	Heels almost touching during walking	= 1	
	Gait score		
	Balance + Gait score		

The Chester Step Test Graphical Datasheet and Normative Data

Name: _____ Age: _____ MaxHR: _____ b/m 80% MaxHR: _____ b/m

Heart rate (beats/min)
210
200
190
180
170
160
150
140
130
120
110
100
90
80
70
60

ml $O_2 \cdot kg^{-1} \cdot min^{-1}$ 14 17 20 23 26 29 32 35 38 41 44 47 50 53 56 59 62 65 68 71

Date of test				
Aerobic capacity (ml $O_2 \cdot kg^{-1} \cdot min^{-1}$)				
Fitness rating				

Norms for aerobic capacity ml $O_2 \cdot kg^{-1} \cdot min^{-1}$										
	MALES AGE GROUP					FEMALES AGE GROUP				
Fitness rating	15-19	20-29	30-39	40-49	50+	15-19	20-29	30-39	40-49	50+
Excellent	60+	55+	50+	46+	44+	55+	50+	46+	43+	41+
Good	48-59	44-54	39-49	37-45	35-43	44-54	39-49	35-45	34-42	33-40
Average	39-47	35-43	32-38	30-38	27-34	36-43	32-38	29-34	27-33	26-32
Below average	30-38	28-34	22-31	24-29	22-26	25-35	27-31	24-28	22-26	20-25
Poor	<30	<28	<26	<24	<22	<29	<27	<24	<22	<20

Closed Kinetic Chain Upper Extremity Stability Test

Client's name: _____ DOB: _____

Clinician: _____ Date of injury/surgery: _____

Diagnosis: _____ Ht: _____ in. Wt: _____ lb

PROCEDURE

1. Subject assumes push-up (male) or modified push-up (female) position.

2. Subject has to move both hands back and forth from each line as many times as possible in 15 s. Lines are 3 ft apart.

3. Count the number of lines touched by both hands.

4. Begin with one submaximal warm-up. Repeat three times and average.

5. Normalize score by the following formula:

   - Score = $\dfrac{\text{Average number of lines touched}}{\text{Height (in.)}}$

   - Determine power by using following formula (68% body weight = trunk, head, and arms):

   - Power = $\dfrac{68\% \text{ Weight} \times \text{Average number of lines touched}}{15}$

DATA COLLECTION AREA

DATE OF TEST				
Trial	1	2	3	Mean
Touches				
Score:				
Power:				

NORMATIVE DATA

	Males	Females
Average number of touches	21	23

Upper Extremity Throwing Analysis

I. LATERAL VIEW

A. Preparatory	Yes	No	Position
1. Is patient balanced when the knee reaches max height?			
2. Is the head over the middle of the body (observe and position)?			
3. Is lift from thigh and foot directly under knee?			
4. Is hip rotation closed, one pocket to the hitter?			
5. Is weight transferred forward (lateral move to the plate when stride knee starts to move down and toward plate)?			
6. Is weight transfer not rushed?			
7. Does the driving leg remain constant and not drop before forward propulsion?			

B. Stride			
1. Is stride length less than pitcher's height (82-94%)?			
2. Are the pitcher's head and shoulder over the stride leg at ball release?			
3. Is the stride foot landing neutral (not the heel) between open and closed? (−10° to +12°)?			
4. Is the knee flexion angle between 34° and 56° at foot contact?			
5. Do the hips rotate first, followed by the shoulders, after foot contact?			
6. Are the hips, shoulders, and eyes level horizontally?			
7. What is the arm position at foot contact?			
8. Shoulder abduction 89°-110°?			
9. Shoulder external rotation 40°-92°?			
10. Elbow flexion 76°-113°?			

(continued)

(continued from previous page)

C. Cocking	Yes	No	Position
1. Is the cocking position of the hand high and back?	_____	_____	_____
2. Is the external rotation of the shoulder between 84° and 103° (back arched)?	_____	_____	_____
3. Is the elbow flexed between 76° and 113°?	_____	_____	_____
4. Is the horizontal adduction between 7° and 22°?	_____	_____	_____

D. Release			
1. Is the release point with a forward flexed angle of trunk?	_____	_____	_____
2. Is the lead knee angle the same as at foot contact (15°-52°)?	_____	_____	_____

E. Follow-Through			
1. Is the lead knee straightened at ball release?	_____	_____	_____

II. FRONT VIEW

A. Preparatory			
1. Are the head and eyes in the proper position?	_____	_____	_____
2. Is the stride foot alignment proper?	_____	_____	_____
3. Are the front shoulder and hip alignment correct?	_____	_____	_____
4. Is the trunk tilt between 47° and 59° above horizontal at ball release?	_____	_____	_____
5. Is the lead arm action correct?	_____	_____	_____
6. Are the following correct at ball release?	_____	_____	_____
Hand and elbow between 12° and 26°?	_____	_____	_____
Shoulder abduction between 89° and 116°?	_____	_____	_____
Glove tucked?	_____	_____	_____

B. Follow-Through	Yes	No	Position
1. Is the throwing shoulder over the lead leg during deceleration?	_____	_____	_____
2. Does the trunk become horizontal?	_____	_____	_____
3. Does the back of the throwing shoulder appear?	_____	_____	_____
4. Does the thrower end in good fielding position?	_____	_____	_____

III. SUMMARY

_____

_____

_____

_____

_____

_____

_____

_____

_____

IV. RECOMMENDATIONS

_____

_____

_____

_____

_____

_____

_____

_____

_____

References

Chapter 1

American Physical Therapy Association. Guide to physical therapist practice, 2nd ed. *Phys Ther* 2001;81:9-746.

Austin GP. Functional testing and return to activity. In: Magee DJ, Zachazewski JA, Quillen WS (eds.), *Scientific Foundations and Principles of Practice in Musculoskeletal Rehabilitation.* St. Louis: Saunders, 2007.

Davies GJ, Zillmer DA. In: Ellenbecker TS (ed.), *Knee Ligament Rehabilitation.* New York: Churchill Livingston, 2000.

Domholdt E. *Physical Therapy Research.* 2nd ed. Philadelphia: Saunders, 2000.

Ebel RL. Estimates of the reliability of ratings. *Psychometrika* 1951;16:407.

Fleiss JL. *The Design and Analysis of Clinical Experiments.* New York: Wiley, 1986.

Fritz JM, Wainner RS. Examining diagnostic tests: An evidence-based perspective. *Phys Ther* 2001;81:1546-1564.

Huijbregts PA. Spinal motion palpation: A review of reliability studies. *J Man Manipul Ther* 2002;10:24-39.

Jaeschke RZ, Meade MO, Guyatt GH, et al. How to use diagnostic test articles in the intensive care unit: Diagnosing weanability using f/Vt. *Crit Care Med* 1997;25:1514-1521.

Laslett M, Williams M. The reliability of selected pain provocation tests for sacroiliac joint pathology. *Spine* 1994;19:1243-1249.

McGinn T, Guyatt G, Wyer P, et al. Users' guides to the medical literature XXII: How to use articles about clinical decision rules. *JAMA* 2000;284:79-84.

Portney L, Watkins MP. *Foundations of Clinical Research: Applications to Practice.* Norwalk, CT: Appleton and Lange, 1993.

Portney LG, Watkins MP. *Foundations of Clinical Research. Applications to Practice.* 2nd ed. Upper Saddle River, NJ: Prentice Hall, 2000.

Rothstein JM, Echternach JL. *Primer on Measurement: An Introductory Guide to Measurement Issues.* Alexandria, VA: American Physical Therapy Association, 1999.

Sackett DL. A primer on the precision and accuracy of the clinical examination. *JAMA* 1992;267:2638-2644.

Sackett DL, Haynes RN, Guyatt GH, Tugwell P. *Clinical Epidemiology: A Basic Science for Clinical Medicine.* 2nd ed. Boston: Little, Brown, 1992.

Sackett DL, Rosenberg WM, Gray JA, et al. Evidence based medicine: What it is and what it isn't. *BMJ* 1996;312:71-72.

Sackett DL, Straws SE, Richardson WS, Rosenberg W, Haynes RB. *Evidence-Based Medicine: How to Practice and Teach EBM.* 2nd ed. London: Harcourt, 2000.

Task Force on Standards for Measurement in Physical Therapy. Standards for tests and measures in physical therapy practice. *Phys Ther* 1991;71:589-622.

Verbrugge LM, Jette AM. The disablement process. *Soc Sci Med* 1994;38(1):1-14.

Chapter 2

Abernethy PJ. Influence of acute endurance activity on isokinetic strength. *J Strength Cond Res* 1993;7:141-146.

American College of Sports Medicine. *ACSM's Guidelines for Exercise Testing and Prescription.* 7th ed. Philadelphia: Lippincott Williams & Wilkins, 2006.

Anderson B, Burke ER. Scientific, medical, and practical aspects of stretching. *Clin Sports Med* 1991;10:63-86.

Austin GP. Functional testing and return to activity. In: Magee DJ, Zachazewski JE, Quillen WS (eds.), *Scientific Foundations and Principles of Practice in Musculoskeletal Rehabilitation.* St. Louis: Saunders, 2007.

Bracko MR. Can stretching prior to exercise and sports improve performance and prevent injury? *ACSM's Health Fitness J* 2002;6(5):17-22.

Chiu LZF, Barnes JL. The fitness-fatigue model revisited: Implications for planning short-and long-term training. *Strength Cond J.* 2003;25(6):42-51.

Church JB, Wiggins MS, Moode FM, Crist R. Effect of warm-up and flexibility treatments on vertical jump performance. *J Strength Cond Res* 2001;15(3):332-336.

Etnyre BR, Abraham LD. H-reflex changes during static stretching and two variations of proprioceptive neuromuscular facilitation techniques. *Electroenceph Clin Neurophysiol* 1986;63:174-179.

Fleck SJ, Kraemer WJ. *Designing Resistance Training Programs.* 3rd ed. Champaign, IL: Human Kinetics, 2004.

Fletcher IM, Jones B. The effects of different warm-up stretch protocols on 20 meter sprint performance in trained rugby union players. *J Strength Cond Res* 2004;18(4):885-888.

Fowles JR, Sale DG, MacDougall JD. Reduced strength after passive stretch of the human plantarflexors. *J Appl Physiol.* 2000;89(3):1179-1188.

Gray SC, Devito G, Nimmo MA. Effect of active warm-up on metabolism prior to and during intense dynamic exercise. *Med Sci Sports Exerc* 2002;34(12):2091-2096.

Gullich A, Schmidtbleicher D. MVC-induced short-term potentiation of explosive force. *New Stud Athletics* 1996;11:67-81.

Häkkinen K. Neuromuscular fatigue and recovery in male and female athletes during heavy resistance exercise. *Int J Sports Med* 1993;14:53-59.

Harman E. The biomechanics of resistance exercise. In: Baechle TR (ed.), *Essentials of Strength Training and Conditioning*. Champaign, IL: Human Kinetics, 1994.

Heyward VH. *Advanced Fitness and Exercise Prescription*. 5th ed. Champaign, IL: Human Kinetics, 2006.

Hoffman J. *Norms for Fitness, Performance, and Health*. Champaign, IL: Human Kinetics, 2006.

Holt LE, Travis TM, Okita T. Comparative study of three stretching techniques. *Percep Motor Skills* 1970;31:611-616.

Hopkins WG, Schabort EJ, Hawley JA. Reliability of power in physical performance tests. *Sports Med* 2001;31:211-234.

Houston ME, Green HJ, Stull JT. Myosin light chain phosphorylation and isometric twitch potentiation in intact human muscle. *Pflugers Arch* 1985;403:348-352.

Jackson AS, Atkinson G, Hopkins WG. Reliability: A crucial issue for clinicians and researchers. *Med Sci Sports Exerc* 2001;33:S173.

Knudson DV, Magnusson P, McHugh M. Current issues in flexibility fitness. *Phys Fit Sports Res Digest* 2000;3:1-8.

Koch AJ, O'Bryant HS, Stone ME, et al. Effect of warm-up of the standing broad jump in trained and untrained men and women. *J Strength Cond Res* 2003;17(4):710-714.

Kubo K, Kanehisa H, Kawakami Y, Fukunga T. Influence of static stretching on viscoelastic properties of human tendon structures in vivo. *J Appl Physiol* 2001;90:520-527.

Leveritt M, Abernethy PJ. Acute effects of high-intensity endurance exercises on subsequent resistance activity. *J Strength Cond Res* 1999;13:47-51.

Leveritt M, MacLaughlin H, Abernethy PJ. Changes in leg strength 8 and 32 hours after endurance exercise. *J Sport Sci* 2000;18:865-871.

Manske RC, Smith BS, Wyatt FB, et al. Test-retest reliability of lower extremity functional tests after a closed kinetic chain exercise bout. *J Sport Rehabil* 2003;12:119-132.

Pandorf CE, Nindle BC, Montain SJ, Castellani JW, Frykman PN, Leone CD, et al. Reliability assessment of two military relevant occupational physical performance tests. *Can J Appl Physiol* 2003;28:27-37.

Pope RP, Herbert RD, Kirwan JD, Graham BJ. A randomized trial of preexercise stretching for prevention of lower limb injury. *Med Sci Sports Exerc* 2000;32:271-277.

Reiman MP. Training for strength, power, and endurance. In: Manske RC (ed.), *Postoperative Orthopedic Sports Medicine: The Knee and Shoulder*. Philadelphia: Elsevier Science, 2006.

Shrier I. Stretching before exercise: An evidenced based approach. *Br J Sports Med* 2000;34:324-325.

Shrier I. Stretching before exercise does not reduce the risk of local muscle injury: A critical review of the clinical and basic science literature. *Clin J Sports Med* 1999;9:221-227.

Shrier I, Gossal K. Myths and truths of stretching: Individualized recommendations for healthy muscles. *Phys Sportsmed* 2000;28:57-63.

Smith JC, Fry AC, Weiss LW, Li Y, Kinzey SJ. The effects of high-intensity exercise on a 10-second sprint cycle test. *J Strength Cond Res* 2001;15:344-348.

Taylor DC, Dalton JD, Seaber AV, Garrett WE. Viscoelastic properties of muscle-tendon units. The biomechanics of stretching. *Am J Sports Med* 1990;18:300-309.

Thacker SB, Gilchrist J, Stroup DF, Kimsey CD. The impact of stretching on sports injury risk: A systematic review of the literature. *Med Sci Sports Exerc* 2004;36:371-378.

Wallin D, Ekblom B, Grahn R, Nordenborg T. Improvement of muscle flexibility. A comparison between two techniques. *Am J Sports Med* 1985;13:263-268.

Weldon SM, Hill RH. The efficacy of stretching for prevention of exercise-related injury: A systematic review of the literature. *Man Ther* 2003;8:141-150.

Young WB, Behm DG. Effects of running, static stretching and practice jumps on explosive force production and jumping performance. *J Sports Med Phys Fitness* 2003;43:21-27.

Young WB, Jenner A, Griffiths K. Acute enhancement of power performance from heavy load squats. *J Strength Cond Res* 1998;12:82-84.

Chapter 3

Ageberg E, Zatterstrom R, Friden T, Moritz U. Individual factors affecting stabilometry and one-leg hop tests in 75 healthy subjects, aged 15-44 years. *Scand J Med Sci Sports* 2001;11:47-53.

Ashby BM, Heegaard JH. Role of arm motion in the standing long jump. *J Biomech* 2002;35:1631-1637.

Barber SD, Noyes FR, Mangine RE, McCloskey JW, Hartman W. Quantitative assessment of functional limitations in normal and anterior cruciate ligament-deficient knees. *Clin Orthop* 1990;204-219.

Cook G, Burton L, Fields K, Kiesel K. *The Functional Movement Screen*. Self-published training manual. Danville, VA, 1998.

Davies GJ, Zillmer DA. Functional progression of a patient through a rehabilitation program. *Orthop Phys Ther Clin N Am* 2000;9(2):103-118.

Flanagan SP, Kulig K. Assessing musculoskeletal performance of the back extensors following a single-level microdiscectomy. *J Orthop Sports Phys Ther* 2007;37:356-363.

Fritz JM, George SZ, Delitto A. The role of fear-avoidance beliefs in acute low back pain: Relationships with current and future disability and work status. *Pain* 2001;94:7-15.

Gribble PA, Hertel J. Considerations for normalization of measures of the Star Excursion Balance Test. *Meas Phys Educ Sci* 2003;7:89-100.

Hertel J, Braham RA, Hale SA, Olmsted-Kramer LC. Simplifying the star excursion balance test: Analyses of subjects with and without chronic ankle instability. *J Orthop Sports Phys Ther* 2006;36:131-137.

Hertel J, Miller SJ, Denegar CR. Intratester and intertester reliability during the star excursion balance test. *J Sport Rehabil* 2000;9:104-116.

Kinzey SJ, Armstrong CW. The reliability of the star excursion test in assessing dynamic balance. *J Orthop Sports Phys Ther* 1998;27:356-360.

Manske RC, (ed). *Postsurgical Orthopedic Sports Rehabilitation: Knee and Shoulder*. St. Louis: Elsevier, 2006.

Manske RC, Andersen J. Test-retest reliability of the lower extremity functional reach test. *J Orthop Sports Phys Ther (Abstract)* 2004;34(1):A52-53.

McGill SM, Childs A, Liebenson C. Endurance times for stabilization exercises: Clinical targets for testing and training from a normal database. *Arch Phys Med Rehabil* 1999;80:941-944.

Noyes FR, Barber SD, Mangine RE. Abnormal lower limb symmetry determined by functional hop tests after anterior cruciate ligament rupture. *Am J Sports Med* 1991;19:513-518.

Olmsted LC, Carcia CR, Hertel J, Shultz SJ. Efficacy of the Star Excursion Balance Tests in detecting reach deficits in subjects with chronic ankle instability. *J Athl Train* 2002;37:501-506.

Plisky PJ, Rauh MJ, Kaminski TW, Underwood FB. Star Excursion Balance Test as a predictor of lower extremity injury in high school basketball players. *J Orthop Sports Phys Ther* 2006;36:911-919.

Reiman MP. Training for strength, power, and endurance. In: Manske RC (ed.), *Postsurgical Orthopedic Sports Rehabilitation: Knee and Shoulder.* St. Louis: Mosby, 2006.

van der Harst JJ, Gokeler A, Hof AL. Leg kinematics and kinetics in landing from single-leg hop for distance. A comparison between dominant and non-dominant leg. *Clin Biomech* 2007;22:674-680.

van Trijffel E, Anderegg Q, Bossuyt PMM, Lucas C. Inter-examiner reliability of passive assessment of intervertebral motion in the cervical and lumbar spine: A systematic review. *Man Ther* 2005;10:256-269.

Chapter 4

Albert WJ, Bonneau J, Stevenson JM, Gledhill N. Back fitness and back health assessment considerations for the Canadian physical activity, fitness and lifestyle appraisal. *Can J Appl Physiol* 2001;26:291-317.

Bray GA, Gray DS. Obesity part I—pathogenesis. *West J Med* 1988;149:432-441.

Forthomme B, Croisier JL, Ciccarone G, Crielaard JM, Cloes M. Factors correlated with volleyball spike velocity. *Am J Sports Med* 2005;33:1513-1519.

Foss ML, Keteyian SJ. Exercise, body composition, and weight control. In: Foss ML, Keteyian SJ (eds.), *Fox's Physiological Basis for Exercise and Sport.* 6th ed. Boston: McGraw-Hill, 1998.

Gross MT, Credle JK, Hopkins LA, Kollins TM. Validity of knee flexion and extension peak torque prediction models. *Phys Ther* 1990;70:3-10.

Gross MT, Dailey ES, Dalton MD, et al. Relationship between lifting capacity and anthropometric measures. *J Orthop Sports Phys Ther* 2000;30:237-247.

Gross MT, McGrain P, Demilio N, Plyler L. Relationship between multiple predictor variables and normal knee torque production. *Phys Ther* 1989;69:54-62.

Heyward VH. *Advanced Fitness and Exercise Prescription.* 5th ed. Champaign, IL: Human Kinetics, 2006.

Malina RM. Anthropometry. In: Maud PJ, Foster C (eds.), *Physiological Assessment of Human Fitness.* Champaign, IL: Human Kinetics, 1995.

Norgan NG. Population differences in body composition in relation to the body mass index. *Eur J Clin Nutr* 1994;48(suppl):S10-S25.

Payne N, Gledhill N, Katzmarzyk PT, Jamnik VK, Keir PJ. Canadian musculoskeletal fitness norms. *Can J Appl Physiol* 2000;25:430-442.

Rudolf MC, Walker J, Cole TJ. What is the best way to measure waist circumference? *Int J Pediatr Obes* 2007;2:58-61.

Smalley KJ, Knerr AN, Kendrick ZV, Colliver JA, Owen OE. Reassessment of body mass indices. *Am J Clin Nutr* 1990;52:405-408.

Smith SC Jr, Haslam D. Abdominal obesity, waist circumference and cardio-metabolic risk: Awareness among primary care physicians, the general population and patients at risk—the Shape of the Nations survey. *Curr Med Res Opin* 2007;23:29-47.

Chapter 5

Bandy WD, Irion JM. The effect of time on static stretch on the flexibility of the hamstring muscles. *Phys Ther* 1994;74:54-61.

Bandy WD, Irion JM, Briggler M. The effect of time and frequency of static stretching on flexibility of the hamstring muscles. *Phys Ther* 1997;77:1090-1096.

Borstad JD. Resting position variables at the shoulder: Evidence to support a posture-improvement association. *Phys Ther* 2006;86:549-557.

Bullock-Saxton J, Bullock M. Repeatability of muscle length measures around the hip. *Physiother Can* 1994;46:105-109.

Corkery M, Briscoe H, Ciccone N, et al. Establishing normal values for lower extremity muscle length in college-age students. *Phys Ther Sport* 2007;8:66-74.

Fishman L, Dombi G, Michaelson C, et al. Piriformis syndrome: Diagnosis, treatment, and outcome—a 10 year study. *Arch Phys Med Rehabil* 2002;83:295-301.

Flynn TW, Whitman J, Magel J. *Orthopedic Manual Physical Therapy Management of the Cervical-Thoracic Spine & Ribcage.* CD-ROM. Manipulations, Inc., 2000.

Gajdosik R, Lusin G. Hamstring muscle tightness: Reliability of an active-knee extension test. *Phys Ther* 1983;63:1085-1088.

Gajdosik R, Rieck MA, Sullivan DK, et al. Comparison of four clinical tests for assessing hamstring muscle length. *J Orthop Sports Phys Ther* 1993;18:614-618.

Gajdosik RL, Sandler MM, Marr HL. Influence of knee positions and gender on the Ober test for length of the iliotibial band. *Clin Biomech* 2003;18:77-79.

Goldspink DF, Cox VM, Smith SK, et al. Muscle growth in response to mechanical stimuli. *Am J Physiol* 1995;268:E288-E297.

Greenman PE. *Principles of Manual Medicine.* 2nd ed. Baltimore: Lippincott Williams & Wilkins, 1996.

Hanton W, Chandler S. Effects of myofascial release leg pull and sagittal plane isometric contract-relax techniques on passive straight leg raise angle. *J Orthop Sports Phys Ther* 1994;20:138-144.

Hartig DE, Henderson JM. Increasing hamstring flexibility decreases lower extremity overuse injuries in military basic trainees. *Am J Sports Med* 1999;27(2):173-176.

Harvey D, Mansfield C. Measuring flexibility for performance and injury prevention. In: Gore CJ (ed.), *Physiological Tests for Elite Athletes.* Champaign, IL: Human Kinetics, 2000.

Heyward VH. *Advanced Fitness and Exercise Prescription.* 5th ed. Champaign, IL: Human Kinetics, 2006.

Hsieh CY, Walker JM, Gillis K. Straight-leg raising test: Comparison of three instruments. *Phys Ther* 1983;63:1439-1433.

Janda V. *Muscle Function Testing*. London: Butterworth, 1983.

Janda V. Muscles and motor control in cervicogenic disorders: Assessment and management. In: Grant R (ed.), *Physical Therapy of the Cervical and Thoracic Spine*. New York: Churchill Livingstone, 1994.

Kendall F, McCreary EK, Provance PG, Rodgers MM, Romani WA. *Muscles: Testing and Function with Posture and Pain*. 5th ed. Baltimore: Lippincott Williams & Wilkins, 2005.

Kroon P, Kruchowsky T. *The Manual Therapy Institute Manual Therapy Program Manual*. San Marcos, Texas, 2006.

Lederman E. *Fundamentals of Manual Therapy: Physiology, Neurology and Psychology*. London: Churchill Livingstone, 1997.

Lee D. *The Pelvic Girdle: An Approach to the Examination and Treatment of the Lumbopelvic-Hip Region*. 3rd ed. London: Churchill Livingstone, 2004.

Leivseth G, Torstensson J, Reikeras O. The effect of passive muscle stretching in osteoarthritis of the hip. *Clin Sci* 1989;76:113-117.

Manske RC, Jones SM, Bryan TL, et al. Interrater and intrarater reliability of upper extremity muscle length testing, and a comparison of upper extremity muscle length in normal males and college baseball pitchers. *J Orthop Sports Phys Ther* 2006;36(1):A83.

Melchione WE, Sullivan MS. Reliability of measurements obtained by use of an instrument designed to indirectly measure iliotibial band length. *J Orthop Sports Phys Ther* 1993;18:511-515.

Reese NB, Bandy WD (eds.). *Joint Range of Motion and Muscle Length Testing*. Philadelphia: Saunders, 2002.

Reese NB, Bandy WD. Use of an inclinometer to measure flexibility of the iliotibial band using the Ober test and the modified Ober test: Differences in magnitude and reliability of measurements. *J Orthop Sports Phys Ther* 2003;33:326-330.

Rose MJ. The statistical analysis of the intra-observer repeatability of four clinical measurement techniques. *Physiotherapy* 1991;77:89-91.

Sahrmann S. *Diagnosis and Treatment of Movement Impairment Syndromes*. St. Louis: Mosby, 2002.

Sapega A, Quedenfeld T, Moyer R, Butler R. Biophysical factors in range-of-motion exercise. *Phys Sportsmed* 1981;9:57-65.

Sullivan MK, Dejulia JJ, Worrell TW. Effect of pelvic position and stretching method on hamstring muscle flexibility. *Med Sci Sports Exerc* 1992;24:1383-1389.

Wang SS, Whitney SL, Burdett RC, et al. Lower extremity muscular flexibility in long distance runners. *J Orthop Sports Phys Ther* 1993;17:102-107.

Webright W, Randolph BJ, Perrin D. Comparison of nonballistic active knee extension in neural slump position and static stretch techniques of hamstring flexibility. *J Orthop Sports Phys Ther* 1997;26:7-13.

Weldon SM, Hill RH. The efficacy of stretching for prevention of exercise-related injury: A systematic review of the literature. *Man Ther* 2003;8:141-150.

Wessling KC, DeVane DA, Hylton CR. Effects of static stretch versus static stretch and ultrasound combined on triceps surae muscle extensibility in healthy women. *Phys Ther* 1987;67(5):674-679.

Worrell TW, Smith TL, Winegardner J. Effect of hamstring stretching on hamstring muscle performance. *J Orthop Sports Phys Ther* 1994;20:154-159.

Yang S, Alnaqeeb M, Simpson H, Goldspink G. Changes in muscle fibre type, muscle mass and IGF-1 gene expression in rabbit skeletal muscle subjected to stretch. *J Anat* 1997;190:613-622.

Youdas JW, Krause DA, Harmsen WS, Laskowski E. The influence of gender and age on hamstring muscle length in healthy adults. *J Orthop Sports Phys Ther* 2005;35:246-252.

Chapter 6

Alaranta H, Hurri H, Heliovaara M, Soukka A, Harju R. Non-dynamometric trunk performance tests: Reliability and normative data. *Scand J Rehabil Med* 1994;26:211-215.

Clark MA. *Integrated Training for the New Millennium*. Thousand Oaks, CA: National Academy of Sports Medicine, 2001.

Cook G, Burton L, Fields K, Kiesel K. *The Functional Movement Screen*. Self-published training manual. Danville, VA, 1998.

Jull GA, Janda V. Muscle and motor control in low back pain. In: Twomey LT, Taylor JR (eds.), *Physical Therapy of the Low Back: Clinics in Physical Therapy*. New York: Churchill Livingstone, 1987.

Kiesel K. The functional movement screen: Predicting injury. Paper presented at the Team Concept Conference, Las Vegas, 2006.

Minick K, Burton L, Kiesel K. A reliability study of the functional movement screen. Paper presented at the National Strength and Conditioning Conference, Atlanta, 2007.

Portney LG, Watkins MP. *Foundation of Clinical Research: Applications to Practice*. 2nd ed. Upper Saddle River, NJ: Prentice Hall, 2000.

Sim J, Wright CC. The kappa statistic in reliability studies: Use, interpretation, and sample size requirements. *Phys Ther* 2005;85(3):257-268.

www.functionalmovement.com

Chapter 7

Ageberg E, Zatterstrom R, Moritz U. Stabilometry and one-leg hop test have high test-retest reliability. *Scand J Med Sci Sports* 1998;8(4):198-202.

Anderson MK, Hall SJ, Martin M. *Sports Injury Management*. 2nd ed. Philadelphia: Lippincott Williams & Wilkins, 2000.

Atwater SW, Crowe TK, Deitz JC, Richardson PK. Interrater and test-retest reliability of two pediatric tests. *Phys Ther* 1990;70(2):79-87.

Austin GP, Scibek JS. Intrarater and interrater reliability during the anterior balance and reach test. *J Orthop Sports Phys Ther* 2002;32(1):A47.

Bohannon R. Standing balance, lower extremity muscle strength, and walking performance of patients referred for physical therapy. *Percept Mot Skills* 1995;80:379-385.

Dite W, Temple VA. A clinical test of stepping and change of direction to identify multiple falling older adults. *Arch Phys Med Rehabil* 2002;83:1566-1571.

Eechaute C, Vaes P, Duquet W. Functional performance deficits in patients with CAI: Validity of the multiple hop test. *Clin J Sport Med* 2008;18:124-129.

Fredericks CM. Basic sensory mechanisms and the somatosensory system: Touch and proprioception. In: Fredericks CM, Saladin LK (eds.), *Pathophysiology of the Motor Systems*, pp. 96-101. Philadelphia: Davis, 1996.

Fujisawa H, Takeda R. A new clinical test of dynamic standing balance in the frontal plane: The side-step test. *Clin Rehabil* 2006;20:340-346.

Gillquist J. Knee ligaments and proprioception. *Acta Orthop Scand* 1996;67:533-535.

Gribble PA, Hertel J. Considerations for normalization of measures of the Star Excursion Balance Test. *Meas Phys Educ Sci* 2003;7:89-100.

Guskiewicz KM, Ross SE, Marshall SW. Postural stabilities and neuropsychological deficits after concussion in collegiate athletes. *J Athl Train* 2001;36:263-273.

Hertel J, Braham RA, Hale SA, Olmsted-Kramer LC. Simplifying the star excursion balance test: Analyses of subjects with and without chronic ankle instability. *J Orthop Sports Phys Ther* 2006;36:131-137.

Hertel J, Miller SJ, Denegar CR. Intratester and intertester reliability during the star excursion balance test. *J Sport Rehabil* 2000;9:104-116.

Johnson BL, Nelson JK. *Practical Measurements for Evaluation in Physical Education*. 5th ed. New York: Macmillan, 1986.

Johnsson E, Henriksson M, Hirschfeld H. Does the functional reach test reflect stability limits in elderly people? *J Rehabil Med* 2002;35:26-30.

Kinzey SJ, Armstrong CW. The reliability of the star excursion test in assessing dynamic balance. *J Orthop Sports Phys Ther* 1998;27:356-360.

Loudon JK, Wiesner D, Goist-Foley HL, et al. Intrarater reliability of functional performance tests for subjects with patellofemoral pain syndrome. *J Athl Train* 2002;37(3):256-261.

Madey SM, Cole JK, Brand RA. The sensory role of the anterior ligament. In: DW Jackson (ed.), *The Anterior Cruciate Ligament: Current and Future Concepts*, pp. 23-33. New York: Raven Press, 1993.

Manske RC, Andersen J. Test-retest reliability of the lower extremity functional reach test. *J Orthop Sports Phys Ther (Abstract)* 2004;34(1):A52-53.

Olmsted LC, Carcia CR, Hertel J, Shultz SJ. Efficacy of the Star Excursion Balance Tests in detecting reach deficits in subjects with chronic ankle instability. *J Athl Train* 2002;37:501-506.

Onate JA, Beck BC, Van Lunen BL. On-field testing environment and balance error scoring system performance during a preseason screening of healthy collegiate baseball players. *J Athl Train* 2007;42:446-451.

Plisky PJ, Rauh MJ, Kaminski TW, Underwood FB. Star Excursion Balance Test as a predictor of lower extremity injury in high school basketball players. *J Orthop Sports Phys Ther* 2006;36:911-919.

Riemann BL, Caggiano NA, Lephart SM. Examination of a clinical method of assessing postural control during a functional performance task. *J Sport Rehabil* 1999;8:171-183.

Riemann BL, Guskiewicz KM. Effects of mild head injury on postural stability as measured through clinical balance testing. *J Athl Train* 2000;35:19-25.

Robinson RH, Gribble PA. Support for a reduction in the number of trials needed for the star excursion balance test. *Arch Phys Med Rehabil* 2008;89:364-370.

Sherrington CS. *The Integrative Action of the Nervous System.* New Haven, CT: Yale University Press, 1906.

Starkey C, Ryan J. *Evaluation of Orthopedic and Athletic Injuries.* 2nd ed. Philadelphia: Davis, 2003.

Tinetti ME. Performance oriented assessment of mobility problems in elderly patients. *JAGS* 1986;34:119-126.

Valovich McLeod TC, Perrin DH, Guskiewicz KM, Shultz SJ, Diamond R, Gansneder BM. Serial administration of clinical concussion assessment and learning effects in healthy young athletes. *Clin J Sports Med* 2004;14:287-295.

Whitney SL, Marchetti GF, Morris LO, Sparto PJ. The reliability and validity of the four square step test for people with balance deficits secondary to a vestibular disorder. *Arch Phys Med Rehabil* 2007;88:99-104.

Chapter 8

American College of Sports Medicine. *Guidelines for Exercise Testing and Prescription*. Philadelphia: Lippincott Williams & Wilkins, 1995.

American Heart Association. Exercise standards: A statement for health professionals from the American Heart Association. *Circulation* 1990;82:2286-2322.

Brewer J, Ramsbottom R, Williams C. *Multi-stage Fitness Test.* Belconnen, Australian Capital Territory: Australian Coaching Council, 1988.

D'Alonzo K, Marbach K, Vincent L. A comparison of field measures to assess cardiorespiratory fitness among neophyte exercisers. *Biol Res Nurs* 2006;8:7-14.

Harman E, Garhammar J, Pandorf C. Administration, scoring, and interpretation of selected tests. In: Baechle TR, Earle RW (eds.), *Essentials of Strength Training and Conditioning*. 2nd ed. Champaign, IL: Human Kinetics, 2000.

Hoffman J. *Norms for Fitness, Performance, and Health.* Champaign, IL: Human Kinetics, 2006.

Hoffman RJ, Collingwood TR. *Fit for Duty.* 2nd ed. Champaign, IL: Human Kinetics, 2005.

Kline CJ, Porcari R, Hintermeister P, et al. Estimation of VO2 max from a one-mile track walk, gender, age and body weight. *Med Sci Sports Exerc* 1987;19:253-259.

Latin RW, Berg K, Baechle T. Physical and performance characteristics of NCAA Division I male basketball players. *J Strength Cond Res* 1994;8:214-218.

Leger L, Gadoury C. Validity of the 20m shuttle run test with 1 minute stages to predict VO2 max in adults. *Can J Sport Sci* 1989;14:21-26.

Leger LA, Lambert J. A maximal multistage 20m shuttle run test to predict VO2 max. *Eur J Appl Phys* 1982;49:1-5.

Leger LA, Mercier D, Gadoury C, Lambert J. The multistage 20 metre shuttle run test for aerobic fitness. *J Sports Sci* 1988;6:93-101.

McArdle WD, Katch FI, Katch VL, (eds). *Exercise Physiology. Energy, Nutrition, and Human Performance.* 5th ed. Lippincott Williams and Wilkins, 2001

McArdle WD, Katch FI, Katch VL. *Exercise Physiology. Energy, Nutrition, and Human Performance,* p. 126. Baltimore: Williams & Wilkins, 1994.

Morrow JR, Jackson A, Disch J, Mood D. *Measurement and Evaluation in Human Performance.* 3rd ed. Champaign, IL: Human Kinetics, 2005.

Physical Fitness Specialist Manual, The. Dallas: Cooper Institute for Aerobics Research, 2005.

Ramsbottom R, Brewer J, Williams C. A progressive shuttle run test to estimate maximal oxygen uptake. *Br J Sports Med* 1988;22:141-145.

Shvartz E, Reibold RC. Aerobic fitness norms for males and females aged 6-75: A review. *Av Space Environ Med* 1990;61:3-11.

Sykes K. Capacity assessment in the workplace: A new step test *J Occup Health* 1995;1:20-22.

Sykes K, Roberts A. The Chester step test—a simple yet effective tool for the prediction of aerobic capacity. *Physiotherapy* 2004;90:183-188.

www.presidentschallenge.org.

Chapter 9

Ageberg E, Zatterstrom R, Moritz U. Stabilometry and one-leg hop test have high test-retest reliability. *Scand J Med Sci Sports* 1998;8(4):198-202.

Ajan T, Baroga L. *Weightlifting: Fitness for All Sports.* Budapest: International Weightlifting Federation, Medicina, 1988.

Alkjaer T, Simonsen EB, Magnusson SP, Aagaard H, Dyhre-Poulsen P. Differences in the movement pattern of a forward lunge in two types of anterior cruciate ligament deficient patients: Copers and non-copers. *Clin Biomech (Bristol, Avon)* 2002;17:586-593.

Andersen M, Foreman T. Return to competition: Functional rehabilitation. In: Zachazewski J, Magee D, Quillen W (eds.), *Athletic Injuries and Rehabilitation,* pp. 229-261. Philadelphia: Saunders, 1996.

Anderson AF. Rating scale. In: Fu FH, Harner CD, Vince KG (eds.), *Knee Surgery,* pp. 275-296. Baltimore: Williams & Wilkins, 1994.

Augustsson J, Thomee R, Karlsson J. Ability of a new hop test to determine functional deficits after anterior cruciate ligament reconstruction. *Knee Surg Sports Traumatol Arthrosc* 2004;12:350-356.

Baechle TR, Earle RW. *Weight Training: A Text Written for the College Student.* Omaha: Creighton University Press, 1989.

Baker D, Nance S, Moore M. The load that maximizes the average mechanical power output during jump squats in power trained athletes. *J Strength Cond Res* 2001;15:92-97.

Bandy WD, Rusche KR, Tekulve FY. Reliability and limb symmetry for five unilateral functional tests of the lower extremities. *Isokin Exerc Sci* 1994;4:108-111.

Barber SD, Noyes FR, Mangine RE, McCloskey JW, Hartman W. Quantitative assessment of functional limitations in normal and anterior cruciate ligament deficient knees. *Clin Orthop* 1990;255:204-214.

Bolgla LA, Keskula DR. Reliability of lower extremity functional performance tests. *J Orthop Sports Phys Ther* 1997;3:138-142.

Booher LD, Hench KM, Worrell TW, Stikeleather J. Reliability of three single-leg hop tests. *J Sport Rehabil* 1993;2:165-170.

Bremander AB, Dahl LL, Roos EM. Validity and reliability of functional performance tests in meniscectomized patients with or without knee osteoarthritis. *Scand J Med Sci Sports* 2007;17:120-127.

Brosky JA, Nitz AJ, Malone TR, Caborn DNM, Rayens MK. Intrarater reliability of selected clinical outcome measures following anterior cruciate ligament reconstruction. *J Orthop Sports Phys Ther* 1999;29(1):39-48.

Brzycki M. Strength testing: Predicting a one-rep max from reps-to-fatigue. *JOHPERD* 1993;64:88-90.

Callan SD, Brunner DM, Devolve KL, et al. Physiological profiles of elite freestyle wrestlers. *J Strength Cond Res* 2000;14:162-169.

Chapman PP, Whitehead JR, Binkert RH. The 225-lb reps-to-fatigue test as a submaximal estimate of 1RM bench press performance in college football players. *J Strength Cond Res* 1998;12:258-261.

Chu DA. *Explosive Power and Strength.* Champaign, IL: Human Kinetics, 1996.

Clark MA. *Integrated Kinetic Chain Assessment.* Thousand Oaks, CA: National Academy of Sports Medicine, 2000.

Clark NC, Gumbrell CJ, Rana S, Traole CM, Morrissey MC. Intra-tester reliability and measurement error of the adapted cross-over hop for distance. *Phys Ther Sport* 2002;3:143-151.

Colby SM, Hintermeister RA, Torry MR, Steadman RJ. Lower limb stability with ACL impairment. *J Orthop Sports Phys Ther* 1999;29(8):444-454.

Considine WJ, Sullivan WJ. Relationship of selected tests of leg strength and leg power in college men. *Res Q* 1973;44:404-416.

Crill MT, Kolba C, Chleboun GS. Using lunge measurements for baseline fitness testing. *J Sport Rehabil* 2004;13:44-53.

Daniel D, Andersen AF. Evaluation of treatment results. In: Finerman GAM, Noyes FR (eds.), *Biology and Biomechanics of the Traumatized Synovial Joint: The Knee as a Model,* pp. 573-584. Rosemont, IL: American Academy of Orthopaedic Surgeons, 1991.

Daniel D, Malcolm L, Stone ML, Perth H, Morgan J, Riehl B. Quantification of knee stability and function. *Contemp Orthop* 1982;5:83-91.

Daniel DM, Stone ML, Riehl B, Moore MR. A measurement of lower limb function. The one leg hop-for-distance. *Am J Knee Surg* 1988;1:212-214.

DeCarlo MS, Sell KE. Normative data for range of motion and single-leg hop in high school athletes. *J Sports Rehabil.* 1997;6:246-255.

Delitto A, Irrgang JJ, Harner CD, Fu FH. Relationship of isokinetic quadriceps peak torque and work to one legged hop and vertical jump in ACL reconstructed subjects [abstract]. *Phys Ther* 1993;73(6):S85.

Dugan SA, Frontera WR. Muscle fatigue and muscle injury. *Phys Med Rehabil Clin N Am* 2000;11:385-403.

Elliot J. Assessing muscle strength isokinetically. *JAMA* 1978;240:2408-2412.

Engstrom B, Gornitzka J, Johansson C, Wredmark T. Knee function after anterior cruciate ligament ruptures treated conservatively. *International Orthop* 1993;17:208-213.

Epley B. *Boyd Epley Workout*. Lincoln: University of Nebraska, 1985.

Feagin JA, Lambert KL, Cunningham RR, et al. Consideration of the anterior cruciate ligament injury in skiing. *Clin Orthop* 1987;216:13-18.

Foran B. *High-Performance Sports Conditioning*. Champaign, IL: Human Kinetics, 2000.

Fox EL, Mathews D. *Interval Training: Conditioning for Sports and General Fitness*. Philadelphia: Saunders, 1974.

Friermood HT. Volleyball Skills Contest for Olympic Development. In: United States Volleyball Association, *Annual Official Volleyball Rules and Reference Guide of the U.S. Volleyball Association. 2004*. Colorado Springs, CO.

Fry AC, Kraemer WJ, Weseman CA, et al. The effects of an off-season strength and conditioning program on starters and non-starters in women's intercollegiate volleyball. *J Appl Sport Sci Res* 1991;5:174-181.

Fry AC, Schilling BK, Staron RS, et al. Muscle fiber characteristics and performance correlates of male Olympic style weightlifters. *J Strength Cond Res* 2003;17:746-754.

Garstecki MA, Latin RW, Cuppertt MM. Comparison of selected physical fitness and performance variables between NCAA Division I and II football players. *J Strength Cond Res* 2004;18:292-297.

Gauffin H, Pettersson G, Tegner Y, Tropp H. Function testing in patients with old rupture of the anterior cruciate ligament. *Int J Sports Med* 1990;11:73-77.

Gauffin H, Tropp H. Altered movement and muscular-activation patterns during the one-legged jump in patients with an old anterior cruciate ligament rupture. *Am J Sports Med* 1992;20:182-192.

Gaunt BW, Curd DT. Anthropometric and demographic factors affecting distance hopped and limb symmetry index for the crossover hop-for-distance test in high school athletes. *J Orthop Sports Phys Ther* 2001;31:145-151.

Giorgi A, Wilson GJ, Weatherby RP, Murphy AJ. Functional isometric weight training: Its effects on the development of muscular function and the endocrine system over an 8-week training period. *J Strength Cond Res* 1998;12:18-25.

Glencross DJ. The nature of the vertical jump test and the standing broad jump. *Res Q* 1966;37:353-359.

Goh S, Boyle J. Self evaluation and functional testing two to four years post ACL reconstruction. *Aust Physiol* 1997;43(4):255-262.

Gray G. *Total Body Functional Profile*. Adrian, MI: Wynn Marketing, 2001.

Greenberger HB, Paterno MV. Relationship of knee extensor strength and hopping test performance in the assessment of lower extremity function. *J Orthop Sports Phys Ther* 1995;22(5):202-206.

Greenberger HB, Paterno MV. The test-retest reliability of a one-legged hop for distance in health young adults [abstract]. *J Orthop Sports Phys Ther* 1994;1:62.

Häkkinen KP, Komi V, Alen M. Effective use of explosive type strength training on isometric force and relaxation time, electromyography and muscle fibre characteristics of leg extensor muscles. *Acta Physiol Scand* 1985;125:587-600.

Harman E, Garhammer J, Pandorf C. Administration, scoring, and interpretation of selected tests. In: Baechle TR, Earle RW (eds.), *Essentials of Strength Training and Conditioning*. Champaign, IL: Human Kinetics, 2000.

Harman E, Rosenstein MT, Frykman PN, Rosenstein RM. The effects of arms and countermovement on vertical jumping. *Med Sci Sports Exerc* 1990;22:825-833.

Heyward VH. *Advanced Fitness Assessment and Exercise Prescription*. 5th ed. Champaign, IL: Human Kinetics, 2005.

Hickson RC, Hidaka K, Foster C, Falduto MT, Chatterton RT. Successive time courses of strength development and steroid hormone responses to heavy-resistance training. *J Appl Physiol* 1994;76:663-670.

Hoeger WWK, Barette SL, Hale DF, Hopkins DR. Relationship between repetitions and selected percentages of one repetition maximum. *J Appl Sport Sci Res* 1987;1:11-13.

Hoeger WWK, Hopkins DR, Barette SL, Hale DF. Relationship between repetitions and selected percentages of one repetition maximum: A comparison between untrained and trained males and females. *J Appl Sport Sci Res* 1990;4:47-54.

Hoff J, Almasbakk B. The effects of maximum strength training on throwing velocity and muscle strength in female team-handball players. *J Strength Cond Res* 1995;9:255-258.

Hoffman J. *Norms for Fitness, Performance, and Health*. Champaign, IL: Human Kinetics, 2006.

Hoffman JR, Fry AC, Howard R, Maresh CM, Kraemer WJ. Strength, speed, and endurance changes during the course of a division I basketball season. *J Appl Sport Sci Res* 1991;5:144-149.

Hoffman JR, Tennenbaum G, Maresh CM, Kraemer WJ. Relationship between athletic performance tests and playing time in elite college basketball players. *J Strength Cond Res* 1996;10:67-71.

Hu HS, Whitney SL, Irrgang J, Janosky J. Test-retest reliability of the one-legged vertical jump test and the one-legged standing hop test [abstract]. *J Orthop Sports Phys Ther* 1992;15(1):51.

Itoh H, Ichihashi N, Sakamoto T. Functional test for the knee joint. *Bull Allied Med Sci Kobe* 1989;5:75-81.

Itoh H, Kurosaka M, Yoshiya S, Ichihashi N, Mizuno K. Evaluation of functional deficits determined by four different hop tests in patients with anterior cruciate ligament deficiency. *Knee Surg Sports Traumatol Arthrosc* 1998;6:241-245.

Johnson B. *Practical Measurements for Evaluation in Physical Education*. Edina, MN: Burgess, 1986.

Johnson BL, Nelson JK. *Practical Measurements for Evaluation in Physical Education*. 4th ed. New York: Macmillan College, 1986.

Juris PM, Phillips EM, Dalpe C, Edwards C, Gotlin RS, Kane DJ. A dynamic test of lower extremity function following anterior cruciate ligament reconstruction and rehabilitation. *J Orthop Sports Phys Ther* 1997;26:184-191.

Keays SL, Bullock-Saxton J, Keays AC, Newcombe P. Muscle strength and function before and after anterior cruciate ligament reconstruction using semitendonosus and gracilis. *Knee* 2001;8:229-234.

Kier PJ, Jamnik VK, Glendhill N. Technical-methodological report: A nomogram for peak leg power output in the vertical jump. *J Strength Cond Res* 2003;17:701-703.

Kirkendall DT. Physiology of soccer. In: Garrett WE, Kirkendall DT (eds.), *Exercise and Sport Science.* Philadelphia: Lippincott Williams & Wilkins, 2000.

Knuttgen HG, Kraemer WJ. Terminology and measurement in exercise performance. *J Appl Sport Sci Res* 1987;1:1-10.

Koch AJ, O'Bryant HS, Stone MS, et al. Effect of warm-up on the standing broad jump in trained and untrained men and women. *J Strength Cond Res* 2003;17:710-714.

Kraemer WJ, Fry AC. Strength testing: Development and evaluation of methodology. In: Maud PJ, Foster C (eds.), *Physiological Assessment of Human Fitness.* Champaign, IL: Human Kinetics, 1995.

Kraemer WJ, Gotshalk LA. Physiology of American football. In: Garrett WE, Kirkendall DT (eds.), *Exercise and Sport Science.* Philadelphia: Lippincott Williams & Wilkins, 2000.

Kraemer WJ, Ratamess N, Fry AC, et al. Influence of resistance training volume and periodization on physiological and performance adaptations in college women tennis players. *Am J Sport Med* 2000;28:626-633.

Kramer JF, Nusca D, Fowler P, Webster-Bogaert S. Test-retest reliability of the one-leg hop test following ACL reconstruction. *Clin J Sport Med* 1992;2(4):240-243.

Kuramoto AK, Payne VG. Predicting muscular strength in women: A preliminary study. *Res Q Exerc Sport* 1995;66:168-172.

LaMonte MJ, McKinney JT, Quinn SM, Bainbridge CN, Eisenman PA. Comparison of physical and physiological variables for female college basketball players. *J Strength Cond Res* 1999;13:264-270.

Lander J. Maximum based on reps. *NSCA J* 1984;6:60-61.

Latin RW, Berg K, Baechle T. Physical and performance characteristics of NCAA Division I male basketball players. *J Strength Cond Res* 1994;8:214-218.

Lephart SM, Kocher MS, Harner CD, Fu FH. Quadriceps strength and functional capacity after anterior cruciate ligament reconstruction. Patellar tendon autograft versus allograft. *Am J Sports Med* 1993;21(5):738-743.

Lephart SM, Perrin DH, Fu FH, Gieck JH, McCue FC, Irrgang JJ. Relationship between selected physical characteristics and functional capacity in the anterior cruciate ligament-insufficient athlete. *J Orthop Sports Phys Ther* 1992;16(4):174-181.

Lephart SM, Perrin DH, Fu FH, Minger K. Functional performance tests for the anterior cruciate ligament insufficient athlete. *Athletic Train* 1991;26:44-50.

Lephart SM, Perrin DH, Minger K, Fu F, Gieck GH. Sports specific functional performance tests for the ACL insufficient athlete [abstract]. *Athletic Train* 1988;24:119.

Luthanen P, Komi PV. Segmental contribution to forces in vertical jump. *Eur J Appl Physiol* 1978;38:181-188.

Manske RC, Smith BS, Wyatt F. Test retest reliability of lower extremity functional tests after a closed kinetic chain isokinetic testing bout. *J Sport Rehabil* 2003;12:119-132.

Markovic G, Dizdar D, Jukic I, Cardinale M. Reliability and factorial validity of squat and countermovement jump tests. *J Strength Cond Res* 2004;18(3):551-555.

Mattacola CG, Jacobs CA, Rund MA, Johnson DL. Functional assessment using the step-up-and-over test and forward lunge following ACL reconstruction. *Orthopedics* 2004;27(6):602-608.

Mayhew JL, Ball TE, Arnold ME, Bowen JC. Relative muscular endurance performance as a predictor of bench press strength in college men and women. *J Appl Sport Sci Res* 1992;6:200-206.

McBride JM, Triplett-McBride T, Davie A, Newton RU. The effect of heavy- vs. light-load jump squats on the development of strength, power, and speed. *J Strength Cond Res* 2002;16:75-82.

McCurdy K, Langford G. Comparison of unilateral squat strength between the dominant and non-dominant leg in men and women. *J Sports Sci Med* 2005;4:153-159.

McCurdy K, Langford G, Cline A, Doscheer M, Hoff R. The reliability of 1- and 3 RM tests of unilateral strength in trained and untrained men and women. *J Sports Sci Med* 2004;3:190-196.

McNair PJ, Marshall RN. Landing characteristics in subjects with normal and anterior cruciate ligament deficient knee joints. *Arch Phys Med Rehabil* 1994;75:584-589.

Meir R, Newton R, Curtis E, et al. Physical fitness qualities of professional rugby league football players: Determination of positional differences. *J Strength Cond Res* 2001;15:450-458.

Morales J, Sobonya S. Use of submaximal repetition tests for predicting 1-RM strength in class athletes. *J Strength Cond Res* 1996;10:186-189.

Munich H, Cipriani D, Hall C, Nelson D, Falkel J. The test-retest reliability of an inclined squat strength test protocol. *J Orthop Sports Phys Ther* 1997;26(4):209-213.

Negrete R, Brophy J. The relationship between isokinetic open and closed chain lower extremity strength and functional performance. *J Sport Rehabil* 2003;12:119-132.

Noyes FR, Barber SD, Mangine RE. Abnormal lower limb symmetry determined by function hop tests after anterior cruciate ligament rupture. *Am J Sports Med* 1991;19:513-518.

Ortiz A, Olson S, Roddey TS, Morales J. Reliability of selected physical performance tests in young adult women. *J Strength Cond Res* 2005;19:39-44.

Ostenberg A, Roos H. Injury risk factors in female European football. A prospective study of 123 players during one season. *Scand J Med Sci Sports* 2000;10:279-285.

Pauole K, Madole K, Garhammer J, Lacourse M, Rozenek R. Reliability and validity of the T-test as a measure of agility, leg power, and leg speed in college-aged men and women. *J Strength Cond Res* 2000;14:443-450.

Payne N, Gledhill N, Katzmarzyk PT, Jamnik VK, Keir PJ. Canadian musculoskeletal fitness norms. *Can J Appl Physiol* 2000;25:430-442.

Petschnig R, Baron R, Albrecht M. The relationship between isokinetic quadriceps strength test and hop tests for distance and one-legged vertical jump test following anterior cruciate ligament reconstruction. *J Orthop Sports Phys Ther* 1998;28(1):23-31.

Pfeifer K, Banzer W. Motor performance in different dynamic tests in knee rehabilitation. *Scand J Med Sci Sports* 1999;9:19-27.

Physical Fitness Specialist Certification Manual, The. Dallas: Cooper Institute for Aerobics Research, rev. 1997.

Reid A, Birmingham TB, Stratford PW, Alcock GK, Giffin JR. Hop testing provides a reliable and valid outcome measure during rehabilitation after anterior cruciate ligament reconstruction. *Phys Ther* 2007;87:337-349.

Rhea MR, Ball SD, Phillips WT, Burkett LN. A comparison of linear and daily undulating periodized programs with equated volume and intensity for strength. *J Strength Cond Res* 2002;16:250-255.

Riseberg MA, Ekeland A. Assessment of functional tests after anterior cruciate ligament surgery. *J Orthop Sports Phys Ther* 1994;19(4):212-217.

Robertson DG, Fleming D. Kinetics of standing broad and vertical jumping. *Can J Sports Sci* 1987;12:19-23.

Roos EM, Ostenberg A, Roos H, Ekdahl C, Lohmander LS. Long-term outcome of meniscectomy: Symptoms, function, and performance tests in patients with or without radiographic osteoarthritis compared to matched controls. *Osteoarthritis Cart* 2001;9:316-324.

Ross MD, Langford B, Whelan PJ. Test-retest reliability of 4 single-leg horizontal hop tests. *J Strength Cond Res* 2002;16(4):617-622.

Sanborn K, Boros R, Hruby J, et al. Short-term performance effects of weight training with multiple sets not to failure vs a single set to failure in women. *J Strength Cond Res* 2000;14:328-331.

Sekiya I, Muneta T, Ogiuchi T, Yagishita K, Yamamoto H. Significance of the single-legged hop test to the anterior cruciate ligament-reconstructed knee in relation to muscle strength and anterior laxity. *Am J Sports Med* 1998;26:384-388.

Seminick DM. Testing protocols and procedures. In: Baechle TR (ed.), *Essentials of Strength Training and Conditioning*. Champaign, IL: Human Kinetics, 1994.

Sernert N, Kartus J, Kohler K, Stener S, Larsson J, Eriksson BI, Karlsson J. Analysis of subjective, objective and functional examination tests after anterior cruciate ligament reconstruction. A follow-up of 527 patients. *Knee Surg Sports Traumatol Arthrosc* 1999;7:160-165.

Sewell LP, Lander JE. The effects of rest on maximal efforts in the squat and bench press. *J Appl Sport Sci Res* 1991;5:96-99.

Shetty AB, Etnyre BR. Contribution of arm movement to the force components of a maximum vertical jump. *J Orthop Sports Phys Ther* 1989;11:198-201.

Smith HK, Thomas SG. Physiological characteristics of elite female basketball players. *Can J Sport Sci* 1991;16:289-295.

Soares J, Mendes OC, Neto CB, Matsudo VKR. Physical fitness characteristics of Brazilian national basketball team as related to game functions. In: Day JAP (ed.), *Perspectives in Kinanthropometry*. Champaign, IL: Human Kinetics, 1986.

Stockbrugger BA, Haennel RG. Contributing factors to performance of a medicine ball explosive power test: A comparison between jump and nonjump athletes. *J Strength Cond Res* 2003;17:768-774.

Stone MH, O'Bryant HS. *Weight Training: A Scientific Approach*. Edina, MN. Burgess International Group, 1987.

Stone MH, O'Bryant HS, McCoy L, Coglianese R, Lehmkuhl M, Schilling B. Power and maximum strength relationships during performance of dynamic and static weighted jumps. *J Strength Cond Res* 2003;17:140-147.

Swarup M, Irrgang JJ, Lephart S. Relationship of isokinetic quadriceps peak torque and work to one legged hop and vertical jump [abstract]. *Phys Ther* 1992;72(6):S88.

Tegner Y, Lysholm J, Lysholm M, Gillquist J. A performance test to monitor rehabilitation and evaluate anterior cruciate ligament injuries. *Am J Sports Med* 1986;14:156-159.

Toji H, Suei K, Kaneko M. Effects of combining training loads on relations among force, velocity, and power development. *Can J Appl Physiol* 1997;22:328-336.

Unger CL, Wooden MMJ. Effect of foot intrinsic muscle strength training on jump performance. *J Strength Cond Res* 2000;14(4):373-378.

Wathen D. In: Baechle TR (ed.), *Essentials of Strength Training and Conditioning*. Champaign, IL: Human Kinetics, 1994.

Wiklander J, Lysholm J. Simple tests for surveying muscle strength and muscle stiffness in sportsmen. *Int J Sports Med* 1987;8(1):50-54.

Wilk KE, Romaniello WT, Soscia SM, Arrigo CA, Andrews JR. The relationship between subjective knee scores, isokinetic testing, and functional testing in the ACL reconstructed knee. *J Orthop Sports Phys Ther* 1994;20(2):60-73.

Wisloff U, Castagna C, Helgerud J, Jones R, Hoff J. Strong correlation of maximal squat strength with sprint performance and vertical jump height in elite soccer players. *Br J Sports Med* 2004;38:285-288.

Wisloff U, Helgerud J, Hoff J. Strength and endurance of elite soccer players. *Med Sci Sports Exerc* 1998;30:462-467.

Woolstenhulme MT, Kerbs Bailey B, Allsen P. Vertical jump, anaerobic power, and shooting accuracy are not altered 6 hours after strength training in collegiate women basketball players. *J Strength Cond Res* 2004;18:422-425.

Young WB, MacDonald C, Flowers MA. Validity of double- and single-leg vertical jumps as tests of leg extensor muscle function. *J Strength Cond Res* 2001;15:6-11.

Chapter 10

Alricsson M, Harms-Ringdahl K, Werner S. Reliability of sports related functional tests with emphasis on speed and agility in young athletes. *Scand J Sci Sports* 2001;11:229-232.

Baechle TR, Earle RW (eds.). *Essentials of Strength Training and Conditioning*. Champaign, IL: Human Kinetics, 2000.

Buckeridge A, Farrow D, Gastin P, McGrath M, Morrow P, Quinn A, et al. Protocols for the physiological assessment of high-performance tennis players. In: Gore CJ (ed.), *Physiological Tests for Elite Athletes*. Champaign, IL: Human Kinetics, 2000.

Ellis L, Gastin P, Lawrence S, Savage B, Buckeridge A, Stapff A, et al. Protocols for the physiological assessment of team sport players. In: Gore CJ (ed.), *Physiological Tests for Elite Athletes*. Champaign, IL: Human Kinetics, 2000.

Fry AC, Kraemer WJ, Weseman CA, et al. The effects of an off-season strength and conditioning program on starters and non-starters in women's intercollegiate volleyball. *J Appl Sport Sci Res* 1991;5:174-181.

Harman E, Garhammer J, Pandorf C. Administration, scoring, and interpretation of selected tests. In: Baechle TR, Earle RW (eds.), *Essentials of Strength Training and Conditioning*. Champaign, IL: Human Kinetics, 2000.

Hoffman J. *Norms for Fitness, Performance, and Health*. Champaign, IL: Human Kinetics, 2006.

Latin RW, Berg K, Baechle T. Physical and performance characteristics of NCAA Division I male basketball players. *J Strength Cond Res* 1994;8:214-218.

McGee KJ, Burkett LN. The National Football League combine: A reliable predictor of draft status? *J Strength Cond Res* 2003;17:6-11.

Ortiz A, Olson SL, Roddey TS, Morales J. Reliability of selected physical performance tests in young adult women. *J Strength Cond Res* 2005;19:39-44.

Pauole K, Madole K, Garhammer J, Lacourse M, Rozenek R. Reliability and validity of the T-test as a measure of agility, leg power, and leg speed in college-aged men and women. *J Strength Cond Res* 2000;14:443-450.

Sawyer DT, Ostarello JZ, Suess EA, Dempsey M. Relationship between football playing ability and selected performance measures. *J Strength Cond Res* 2002;16:611-616.

Semenick DM. Testing protocols and procedures. In: Baechle TR, Earle RW (eds.), *Essentials of Strength Training and Conditioning*. Champaign, IL: Human Kinetics, 2000.

Seminick D. Tests and measurements: The T-test. *NSCA J* 1990;12:36-37.

Stuempfle KJ, Katch FI, Petrie DF. Body composition relates poorly to performance tests in NCAA Division III football players. *J Strength Cond Res* 2003;17:238-244.

Vanderford ML, Meyers MC, Skelly WA, et al. Physiological and sport-specific skill response of Olympic youth soccer athletes. *J Strength Cond Res* 2004;18:334-342.

Wroble RR, Moxley DP. The effect of winter sports participation on high school football players: Strength, power, agility, and body composition. *J Strength Cond Res* 2001;15:132-135.

Chapter 11

Alaranta H, Hurri H, Heliovaara M, Soukka A, Harju R. Non-dynamometric trunk performance tests: Reliability and normative data. *Scand J Rehabil Med* 1994;26:211-215.

Alaranta H, Luoto S, Heliovaara M, Hurri H. Static back endurance and the risk of low back pain. *Clin Biomech* 1995;10:323-324.

Andersson EA, Ma Z, Thorstensson A. Relative EMG levels in training exercises for abdominal and hip flexor muscles. *Scand J Rehabil Med* 1998;30:175-183.

Ashmen KJ, Swanik CB, Lephart SM. Strength and flexibility characteristics of athletes with chronic low-back pain. *J Sport Rehabil* 1996;5:275-286.

Bankoff AD, Furlani J. Electromyographic study of the rectus abdominis and external oblique muscles during exercises. *Electromyogr Clin Neurophysiol* 1984;24:501-510.

Beckman SM, Buchanan TS. Ankle inversion injury and hyper-mobility: Effect on hip and ankle muscle electromyography onset latency. *Arch Phys Med Rehabil* 1995;76:1138-1143.

Biering-Sorensen F. Physical measurements as risk indicators for low back trouble over a one-year period. *Spine* 1984;9:106-119.

Biering-Sorenson F. Physical measurements as risk indicators for low-back trouble over a 1-year period. *Spine* 1989;14:123-125.

Bullock-Saxton JE. Local sensation changes and altered hip muscle function following severe ankle sprain. *Phys Ther* 1994;74:17-28.

Chan RH. Endurance times of trunk muscles in male intercollegiate rowers in Hong Kong. *Arch Phys Med Rehabil* 2005;86:2009-2012.

Chen L-W, Bih L-I, Ho C-C, et al. Endurance times for trunk-stabilization exercises in healthy women: Comparing 3 kinds of trunk-flexor exercises. *J Sport Rehabil* 2003;12:199-207.

Cleland JA, Childs JD, Fritz JM, Whitman JM. Interrater reliability of the history and physical examination in patients with mechanical neck pain. *Arch Phys Med Rehabil* 2006;87:1388-1395.

Cote P, Cassidy JD, Carrol L. The Saskatchewan health and back pain survey: The prevalence of neck pain and related disability in Saskatchewan adults. *Spine* 1998;23:1689-1698.

Cutter NC, Kevorkian CG. *Handbook of Manual Muscle Testing*. New York: McGraw-Hill, 1999.

Ellis L, Gastin P, Lawrence S, Savage B, Buckeridge A, Stapff A, et al. Protocols for the physiological assessment of team sport players. In: Gore CJ (ed.), *Physiological Tests for Elite Athletes*. Champaign, IL: Human Kinetics, 2000.

Falla D, Campbell C, Fagan A, Thompson D, Jull G. An investigation of the relationship between upper cervical flexion range of motion and pressure change during the cranio-cervical flexion test. *Man Ther* 2003;8:92-96.

Faulkner RA, Springings EJ, McQuarrie A, Bell RD. A partial curl-up protocol for adults based on an analysis of two procedures. *Can J Sport Sci* 1989;14:135-141.

Flint MM. An electromyographic comparison of the function of the iliacus and the rectus abdominis muscles. *J Amer Phys Ther Assoc* 1965;45:248-253.

Gilleard WL, Brown JM. An electromyographic validation of an abdominal muscle test. *Arch Phys Med Rehabil* 1994;75:1002-1007.

Gouttebarge V, Wind H, Kuijer PP, Sluiter JK, Frings-Dresen MH. Reliability and agreement of 5 Ergo-Kit functional capacity evaluation lifting tests in subjects with low back pain. *Arch Phys Med Rehabil* 2006;87:1365-1370.

Gracovetsky S, Farfan H. The optimum spine. *Spine* 1986;11:543-572.

Greene WB, KeHaven KE, Johnson TR, et al. (eds.). *Essentials of Musculoskeletal Care*. 2nd ed. Rosemont, IL: American Academy of Orthopedic Surgeons, 2002.

Grimmer K. Measuring the endurance capacity of the cervical short flexor muscle group. *Aust J Physiother* 1994;40:251-254.

Gross DP, Battié MC. Reliability of safe maximum lifting determinations of a functional capacity evaluation. *Phys Ther* 2002;82:364-371.

Gross DP, Battié MC, Cassidy JD. The prognostic value of functional capacity evaluation in patients with chronic low back pain: Part 1: Timely return to work. *Spine* 2004;29:914-919.

Guimaraes AC, Vaz MA, De Compos MI, Marantes R. The contribution of the rectus abdominis and rectus femoris in twelve selected abdominal exercises. An electromyographic study. *J Sports Med Phys Fitness* 1991;31:222-230.

Harman E, Garhammar J, Pandorf C. Administration, scoring, and interpretation of selected tests. In: Baechle TR, Earle RW (eds.), *Essentials of Strength Training and Conditioning*. 2nd ed. Champaign, IL: Human Kinetics, 2000.

Harris KD, Heer DM, Roy TC, Santos DM, Whitman JM, Wainner RS. Reliability of a measurement of neck flexor muscle endurance. *Phys Ther* 2005;85:1349-1355.

Hodges PW, Richardson CA. Contraction of the abdominal muscle associated with movement of the lower limb. *Phys Ther* 1997;77:132-144.

Holmstrom EB, Lindell J, Moritz U. Low back and neck/shoulder pain in construction workers: Occupational workload and psychological risk factors: Part 2 relationship to neck and shoulder pain. *Spine* 1992;17:672-677.

Jager M, Seller K, Raab, Krauspe R, Wild A. Clinical outcome in monosegmental fusion of degenerative lumbar instabilities: Instrumented versus non-instrumented. *Med Sci Mon* 2003;9:CR324-327.

Janda V. *Muscle Function Testing.* London: Butterworth, 1983.

Jull GA. Deep cervical neck flexor dysfunction in whiplash. *J Musculoskel Pain* 2000;8:143-154.

Jull G, Barrett C, Magee R, Ho P. Further characterisation of muscle dysfunction in cervical headache. *Cephalalgia* 1999;19:179-185.

Jull G, Kristjansson E, Dall Alba P. Impairment in the cervical flexors: A comparison of whiplash and insidious onset neck pain patients. *Man Ther* 2004;9:89-94.

Keely G. Posture, body mechanics, and spinal stabilization. In: Bandy WD, Sanders B (eds.), *Therapeutic Exercise: Techniques for Intervention.* Baltimore: Lippincott Williams & Wilkins, 2001.

Kendall FP. *Muscles: Testing and Function.* 4th ed. Baltimore: Williams & Wilkins, 1993.

Knapik JJ, Gerber J. The influence of physical fitness training on the manual material-handling capability and road-marching performance of female soldiers. *US Army Research Laboratory Technical Report ARL-TR-1064.* Aberdeen Proving Ground, MD, 1996.

Krause DA, Youdas JW, Hollman JH, Smith J. Abdominal muscle performance as measured by the double leg-lowering test. *Arch Phys Med Rehabil* 2005;86:1345-1348.

Ladeira CE, Hess LW, Galin BM, Fradera S, Harkness MA. Validation of an abdominal muscle strength test with dynamometry. *J Strength Cond Res* 2005;19:925-930.

Lanning CL, Uhl TL, Ingram CL, Mattacola CG, English T, Newsom S. Baseline values of trunk endurance and hip strength in collegiate athletes. *J Athl Train* 2006;41:427-434.

Leetun DT, Ireland ML, Willson JD, Ballantyne BT, Davis IM. Core stability measures as risk factors for lower extremity injury in athletes. Med Sci Sports Exerc. 2004;36:926-934.

MacDonald DA, Moseley GL, Hodges PW. The lumbar multifidus: Does the evidence support clinical beliefs? *Man Ther* 2006;11:254-263.

Mayoux-Benhamou MA, Revel M, Vallee C. Selective electromyography of dorsal neck muscles in humans. *Exp Brain Res* 1997;113:353-360.

Mayoux-Benhamou MA, Revel M, Vallee C, et al. Longus colli has a postural function on cervical curvature. *Surg Radiol Anat* 1994;16:367-371.

McGill SM. Low back exercises: Prescription for the healthy back and when recovering from injury. In: *Resources Manual for Guidelines for Exercise Testing and Prescription.* 3rd ed. Indianapolis: American College of Sports Medicine; Baltimore: Williams & Wilkins, 1998.

McGill SM, Childs A, Liebenson C. Endurance times for stabilization exercises: Clinical targets for testing and training from a normal database. *Arch Phys Med Rehabil* 1999;80:941-944.

McGill SM, Grenier S, Bavcic N, Cholewicki J. Coordination of muscle activity to assure stability of the lumbar spine. *J Electromyogr Kinesiol* 2003;13:353-359.

Moreland J, Finch E, Stratford P, Balsor B, Gill C. Interrater reliability of six tests of trunk muscle function and endurance. *J Orthop Sports Phys Ther* 1997;26:200-208.

Murray KJ. Hypermobility disorders in children and adolescents. *Best Pract Res Clin Rheumatol* 2006;20:329-351.

Nadler SF, Malanga GA, DePrince M, Stitik TP, Feinberg JH. The relationship between lower extremity injury, low back pain, and hip muscle strength in male and female collegiate athletes. *Clin J Sport Med* 2000;10:89-97.

Novy DM, Simmonds MJ, Lee CE. Physical performance tasks: What are the underlying constructs? *Arch Phys Med Rehabil* 2002;83:44-47.

Pandorf CE, Nindle BC, Montain SJ, Castellani JW, Frykman PN, Leone CD, et al. Reliability assessment of two military relevant occupational physical performance tests. *Can J Appl Physiol* 2003;28:27-37.

Placzek JD, Pagett BT, Roubal PJ, et al. The influence of the cervical spine on chronic headache in women: A pilot study. *J Man Manipul Ther* 1999;7:33-39.

Reese NB. *Muscle and Sensory Testing.* Philadelphia: Saunders, 1999.

Reiman MP, Krier AD, Nelson JA, Rogers MA, Stuke ZO, Smith BS. Reliability of trunk endurance testing modifications. *J Strength Cond Res,* 2009.

Reiman MP, Nelson J, Rogers M, Stuke Z, Zachgo A. Endurance times of trunk muscles in high school weightlifting participants [abstract]. *J Man Manipul Ther* 2006;14:179-180.

Renkawitz T, Boluki D, Grifka J. The association of low back pain, neuromuscular imbalance and trunk extension strength in athletes. *Spine J* 2006;6:673-683.

Richardson C, Jull G, Hodges P, Hides J. *Therapeutic Exercise for Spinal Segmental Stabilization in Low Back Pain: Scientific Basis and Clinical Approach.* New York: Churchill Livingstone, 1999.

Rissanen A, Alaranta H, Sainio P, Harkonen H. Isokinetic and non-dynamometric tests in low back pain patients related to pain and disability index. *Spine* 1994;17:1963-1967.

Schellenberg KL, Lang JM, Chan KM, Burnham RS. A clinical tool for office assessment of lumbar spine stabilization endurance. *Am J Phys Med Rehabil* 2007;86:1-7.

Schmidt GL, Blanpied PR. Analysis of strength tests and resistive exercises commonly used for low-back disorders. *Spine* 1987;12:1025-1034.

Schmidt GL, Blanpied PR, Anderson MA, White RW. Comparison of clinical and objective methods of assessing trunk muscle strength—an experimental approach. *Spine* 1987;12:1020-1024.

Sharp MA, Harman EA, Boutilier BE, Bovee MW, Kraemer WJ. Progressive resistance training program for improving manual materials handling performance. *Work* 1993;3:62-68.

Shields RK, Heiss DG. An electromyographic comparison of abdominal muscle synergies during curl and double straight leg lowering exercises with control of the pelvic position. *Spine* 1997;22:1873-1879.

Smeets R, Hijdra H, Kester A, Hitters M, Knottnerus J. The usability of six physical performance tasks in a rehabilitation population with chronic low back pain. *Clin Rehabil* 2006;20:989-998.

Smith EB, Rasmussen AA, Lechner DE, Gossman MR, Quintana JB, Grubbs BL. The effects of lumbosacral support belts and abdominal muscle strength on functional lifting ability in healthy women. *Spine* 1996;21:356-366.

Udermann BE, Mayer JM, Graves JE, Murray SR. Quantitative assessment of lumbar paraspinal muscle endurance. *J Athl Train* 2003;38:259-262.

Vernon HT, Aker P, Aramenko M, et al. Evaluation of neck muscle strength with a modified sphygmomanometer dynamometer: Reliability and validity. *J Manipul Physiol Ther* 1992;15:343-349.

Visuri T, Ulaska J, Eskelin M, Pulkkinen P. Narrowing of lumbar spinal canal predicts chronic low back pain more accurately than intervertebral disc degeneration: A magnetic resonance imaging study in young Finnish male conscripts. *Mil Med* 2005;170:926-930.

Vitti M, Fujiwara M, Basmanjian JM, Iida M. The integrated roles of the longus colli and sternocleidomastoid muscles: An electromyographic study. *Anat Rec* 1973;177:471-484.

Watson DH, Trott PH. Cervical headache: An investigation of natural head posture and upper cervical flexor muscle performance. *Cephalalgia* 1993;13:272-284.

Wickenden S, Bates S, Maxwell L. An electromyographic evaluation of upper and lower rectus abdominis during various forms of abdominal exercises. *New Zealand J Physiother* 1992;17-21.

Zannotti CM, Bohannon R, Tiberio D, Dewberry MJ, Murray R. Kinematics of the double-leg-lowering test for abdominal muscle strength. *J Orthop Sports Phys Ther* 2002;32:432-436.

Chapter 12

Ajan T, Baroga L. *Weightlifting: Fitness for All Sports.* Budapest: International Weightlifting Federation, Medicina, 1988.

Aussprung DJ, Aussprung J, Gehri D. *Strength Training Design and New Concepts in Clinical Applications.* Duxbury, MA: Strength and Performance Consultants, 1995.

Collins DR, Hedges PB. *A Comprehensive Guide to Sports Skills Tests and Measurement,* pp. 330-333. Springfield, IL: Charles C Thomas, 1978.

Cronin JB, Owen GJ. Upper-body strength and power assessment in women using a chest pass. *J Strength Cond Res* 2004;18(3):401-404.

Davies GJ, Dickoff-Hoffman S. Neuromuscular testing and rehabilitation of the shoulder complex. *J Orthop Sports Phys Ther* 1993;18(2):449-458.

Duncan MJ, Al-Nakeeb Y. Influence of familiarization on a backward, overhead medicine ball explosive power test. *Res Sports Med* 2005;13:345-352.

Ellenbecker TS, Davies GJ. *Closed Kinetic Chain Exercise. A Comprehensive Guide to Multiple-Joint Exercises.* Champaign, IL: Human Kinetics, 2001.

Ellenbecker TS, Manske R, Davies GJ. Closed kinetic chain testing techniques of the upper extremities. *Orthop Phys Ther Clin N Am* 2000;9:219-229.

Ellenbecker TS, Roetert EP. An isokinetic profile of trunk rotation strength in elite tennis players. *Med Sci Sports Exerc* 2004;36:1959-1963.

Field RW. Control tests for explosive events. *NSCA J* 1989;11:63-64.

Field RW. Explosive power test scores among male and female college athletes. *NSCA J* 1991;13:50.

Gillespie J, Keenum S. A validity and reliability analysis of the seated shot put as a test of power. *J Hum Mvmt Stud* 1987;13:97-105.

Goldbeck TG, Davies GJ. Test-retest reliability of the closed kinetic chain upper extremity stability test: A clinical field test. *J Sport Rehabil* 2000;9:35-45.

Ikeda Y, Kijima K, Kawabata K, Fuchimoto T, Ito A. Relationship between side medicine-ball throw performance and physical ability for male and female athletes. *Eur J Appl Physiol* 2007;99:47-55.

Johnson BL, Nelson JK (eds.). *Practical Measurements for Evaluation in Physical Education.* Minneapolis: Burgess, 1979.

Mayhew JL, Bird M, Cole ML, Kock AJ, Jacques JA, Ware JS, et al. Comparison of the backward overhead medicine ball throw to power production in college football players. *J Strength Cond Res* 2005;19(3):514-518.

Mayhew JL, Bemben MG, Piper FC, Ware JS, Rohrs DM, Bemben DA. Assessing bench press power in college football players: The seated shot put. *J Strength Cond Res* 1993;7(2):95-100.

Mayhew JL, Bemben MG, Rohrs DM, Bemben DA. Specificity among anaerobic power tests in college female athletes. *J Strength Cond Res* 1994;8(1):43-47.

Mayhew JL, Bemben MG, Rohrs DM, Piper FC, Willman MK. Comparison of upper body power in adolescent wrestlers and basketball players. *Pediatric Exerc Sci* 1995;7:422-431.

Mayhew JL, Bemben MG, Rohrs DM, Ware J, Bemben DA. Seated shot put as a measure of upper body power in college males. *J Hum Mvmt Stud* 1991;21:137-148.

Mayhew JL, Ware JS, Johns RA, Bemben MG. Changes in upper body power following heavy-resistance strength training in college men. *Int J Sports Med* 1997;18:516-520.

President's Challenge Physical Activity and Fitness Awards Program, a program of the President's Council on Physical Fitness and Sports, U.S. Department of Health and Human Services. www.presidentschallenge.org.

Roetert P, Ellenbecker T. *Complete Conditioning for Tennis,* pp. 12-22. Champaign, IL: Human Kinetics, 1998.

Roush JR, Kitamura J, Waits MC. Reference values for the closed kinetic chain upper extremity stability test (CKCUEST) for collegiate baseball players. *N Am J Sports Phys Ther* 2008;2(3):159-163.

Salonia MA, Chu DA, Cheifetz PM, Freidhoff GC. Upper body power as measured by medicine ball throw distance and its relationship to class level among 10- and 11-year-old female participants in club gymnastics. *J Strength Cond Res* 2004;18(4):695-702.

Stockbrugger BA, Haennel RG. Contributing factors to performance of medicine ball explosive power test: A comparison between jump and non-jump athletes. *J Strength Cond Res* 2003;17(4):768-774.

Stockbrugger BA, Haennel RG. Validity and reliability of a medicine ball explosive power test. *J Strength Cond Res* 2001;15(4):431-438.

Chapter 13

American Alliance for Health, Physical Education, Recreation and Dance. *Res Q Exerc Sport* 1989;60(2):144-151.

Balciunas M, Stonkus S, Abrantes C, Sampaio J. Long term effects of different training modalities on power, speed, skill and anaerobic capacity in young male basketball players. *J Sport Sci Med* 2006;5:163-170.

Bar-Or O. The Wingate anaerobic test: An update on methodology, reliability, and validity. *Sports Med* 1987;4:381-394.

Bar-Or O, Dotan R, Inbar O. A 30 second all-out ergometer test—its reliability and validity for anaerobic capacity. *Israel J Med Sci* 1977;13:126-130.

Bosco CP, Luhtanen P, Komi PV. A simple method for measurement of mechanical power in jumping. *Eur J Physiol* 1983;50:273-282.

Crielaard JM, Pirnay F. Anaerobic and aerobic power of top athletes. *Eur J Appl Physiol* 1981;47:295-300.

Davies GJ, Zillmer DA. Functional progression of a patient through a rehabilitation program. *Orthop Phys Ther Clin N Am* 2000;9(2):103-118.

Gilliam GM. 300 yard shuttle. *NSCA J* 1983;5:46.

Granier P, Mercier B, Mercier J, Anselme F. Aerobic and anaerobic contribution to Wingate test performance in sprint and middle-distance runners. *Eur J Appl Physiol* 1995;70:58-65.

Harman E, Garhammar J, Pandorf C. Administration, scoring, and interpretation of selected tests. In: Baechle TR, Earle RW (eds.), *Essentials of Strength Training and Conditioning*. 2nd ed. Champaign, IL: Human Kinetics, 2000.

Hoffman J. *Norms for Fitness, Performance, and Health*. Champaign, IL: Human Kinetics, 2006.

Hoffman JR, Cooper J, Wendell M, Im J, Kang J. Forthcoming. Effects of β-hydroxy β-methylbutyrate on power performance and indices of muscle damage and stress during high intensity training. *J Strength Cond Res* 2004;18:747-752.

Hoffman JR, Epstein S, Einbinder M, Weinstein Y. The influence of aerobic capacity on anaerobic performance and recovery indices in basketball players. *J Strength Cond Res* 1999;13:407-411.

Inbar O, Bar-Or, O, Skinner JS. *The Wingate Anaerobic Test*. Champaign, IL: Human Kinetics, 1996.

Kraemer WJ, Häkkinen K, Triplett-McBride T, et al. Physiological changes with periodized resistance training in women tennis players. *Med Sci Sports Exerc* 2003;35:157-168.

LaMonte MJ, McKinney JT, Quinn SM, Bainbridge CN, Eisenman PA. Comparison of physical and physiological variables for female college basketball players. *J Strength Cond Res* 1999;13:264-270.

Mangine RE, Noyes FR, Mullen MP, Baker SD. A physiological profile of elite soccer athletes. *J Orthop Sports Phys Ther* 1990;12:147-152.

Maulder P, Cronin J. Horizontal and vertical jump assessment: Reliability, symmetry, discriminative and predictive ability. *Phys Ther Sport* 2005;6:74-82.

Osbeck JS, Maiorca SN, Rundell KW. Validity of field testing to bobsled start performance. *J Strength Cond Res* 1996;10:239-245.

Sands WA. Olympic preparation camps 2000 physical abilities testing. *Technique* 2000;20(10):6-19.

Sands WA, McNeal JR, Ochi MT, Urbanek TL, Jemni M, Stone MH. Comparison of the Wingate and Bosco anaerobic tests. *J Strength Cond Res* 2004;18(4):810-815.

Seiler S, Taylor M, Diana R, Layes J, Newton P, Brown B. Assessing anaerobic power in collegiate football players. *J Appl Sport Sci Res* 1990;4:9-15.

Sinnett AM, Berg K, Latin RW, Noble JM. The relationship between field tests of anaerobic power and 10-km run performance. *J Strength Cond Res* 2001;15(4):405-412.

Smith DJ, Roberts D. Aerobic, anaerobic and isokinetic measures of elite Canadian male and female speed skaters. *J Appl Sport Sci Res* 1991;5:110-115.

Tabor MA, Davies GJ, Kernozek TW, Negrete RJ, Hudson V. A multicenter study of the test-retest reliability of the lower extremity functional test. *J Sport Rehabil* 2002;11:190-201.

Terbizan DJ, Walders M, Seljevold P, Schweigert DJ. Physiological characteristics of masters women fastpitch softball players. *J Strength Cond Res* 1996;10:157-160.

Vanderford ML, Meyers MC, Skelly WA, Stewart CC, Hamilton KL. Physiological and sport-specific skill response of Olympic youth soccer athletes. *J Strength Cond Res* 2004;18:334-342.

Young W. A simple method for evaluating the strength qualities of the leg extensor muscles and jumping abilities. *Strength Cond Coach* 1995;2(4):5-8.

Zabukovec R, Tiidus PM. Physiological and anthropometric profile of elite kickboxers. *J Strength Cond Res* 1995;9:240-242.

Index

Note: The italicized *f* and *t* following page numbers refer to figures and tables, respectively.

1.5 mile run test
  about, 125
  percentile scores, police recruits, 129*t*
  police department personnel passing scores, 129*t*
1RM back squat, 141–142
1RM bench press
  about, 144–145
  age and gender norms, 171*t*
  football players percentile values, 173*t*
  NCAA female athlete percentile values, 174*t*
  normative values among various athletic populations, 172*t*
1RM leg press
  about, 143
  age and gender norms, 170*t*
1RM lifts, as strength indicators, 131
1RM squat
  football players percentile values, 168*t*
  NCAA female athlete percentile values, 168*t*
  normative values, various athletic populations, 169*t*
  strength assessment with, 23
1RM testing protocol, 104
6 m timed hop, reliability, 186*t*
12-minute run test, 125
20-meter shuttle run test, 123, 124
300-meter sprint, 270
300 shuttle run
  about, 264
  percentile ranks in athletes, 271*t*
505 agility test
  about, 194
  normative data for Australian athletes, 207*t*

A
abdominal dynamic endurance
  about, 226

adjusted mean values for subgroups, 237*t*
abdominal muscles endurance test, 224–225
abdominal stage test
  about, 219
  normative values (soccer), 234*t*
accuracy, of testing, 10
active straight leg raise assessment, 97
adductor brevis, 80*t*, 82*t*, 101*t*
adductor longis, 80*t*, 82*t*, 101*t*
adductor magnus, 80*t*, 82*t*, 101*t*
aerobic capacity testing, 27
Ageberg, E., 22, 105
agility, 191
agonistic-antagonistic muscle relationships, 39–40
Alkjaer, T., 136
Al-Nakeeb, Y., 248
alternate pull-up test, 244
alternate single-leg squat test, 147
anaerobic power testing, 263
anconeus, 78*t*
Anderson, M., 156
anterior cruciate ligament injury, 19–22
anterior scalene, 77*t*, 100*t*
anthropometry, 31
Ashby, B.M., 22
Atwater, S.W., 105
Aussprung, D.J., 246
Austin, G.P., 4

B
backward movement agility test
  about, 200
  tennis players, 208*t*
backward overhead medicine ball throw test, 247–248
balance, testing, 22–23, 103
balance error scoring system, 114
ballistic stretching, 12
Bandy, W.D., 150
Banzer, W., 138
Barber, S.D., 150

baseball team, client tests, 22–23
basketball
  1RM bench normative values, 172*t*
  1RM squat normative values, 169*t*
  player vertical jump percentile rates, 181*t*
  T-test descriptive data, 204*t*
  vertical jump normative values, 185*t*
  Wingate anaerobic power test descriptive data, 272*t*
biceps assessment, 53
biceps brachii, 78*t*
biceps femoris, 79*t*, 82*t*, 101*t*
bilateral lower extremity jump, 164, 190*t*
bobsled, test descriptive data, 272*t*
body mass index
  assessment, 33
  Canadian, normative data by gender, 37*t*
  chart by weight, 36*t*
Borstad, J.D., 50, 51
Bosco test
  about, 267
  average and peak power values, 274*t*
brachilais, 78*t*

C
Canadian body mass index, 37*t*
Canadian vertical jump, normative data, 183*t*
capital and cervical neck flexor muscles, 100*t*
capital flexion, 100*t*
cardiorespiratory fitness classification, $\dot{V}O_2$max, 128*t*
cardiovascular response, to testing, 119
carioca test
  about, 134
  research utilizing, 167*t*

cervical deep flexor muscle assessment, 86
cervical flexion, 100*t*
Chester step test, 122, 279
CKCUTEST. *See* closed kinetic chain upper extremity test (CKCUTEST)
Clark, N.C., 156
client tests
    anterior cruciate ligament repair, 19–22
    in clinical research or laboratory setting, 27
    in occupational setting, 23–26
    in preemployment screening, 26–27
clinical research setting, 27
closed kinetic chain upper extremity test (CKCUTEST)
    about, 253
    descriptive and normative data, 262*t*
    reproducible form, 280
co-contraction test, 132
Colby, S.M., 133, 150
compensatory patterns, 14
Cook, G., 92
coracobrachialis, 78*t*
core training, 209
correlation coefficient, 5–6
craniocervical flexion test, 214
Cronin, J., 268

D
Daniel, D., 149
Davies, G.J., 5, 18, 254
DeCarlo, M.S., 149
deep neck flexor muscles, 233*t*
deep neck flexor test, 213–214
deep squat assessment, 91–92
deltoid, 78*t*
Dickoff-Hoffman, S., 254
double-leg jump for distance, 21
double-leg lowering test
    about, 217–218
    average raw values, 234*t*
    muscle grading, 233*t*
Duncan, M.J., 248
dynamic warm-up, 12–13

E
Edgren side-step test, 195
Ekeland, A., 158, 162
Ellenbecker, T.S., 246
endurance ratios
    among muscle groups studied, 235*t*
    approximations among muscle groups studied, 235*t*
endurance testing
    comparison of normative values, 235*t*

comparison of values for normal subjects, 236*t*
energy, primary sources, 15*t*
Engstrom, B., 150
erector spinae, 79*t*, 100*t*, 102*t*
evidence-based practice, 5–8
exercises, low back testing, 25*t*
extensor dynamic endurance test
    about, 227
    adjusted mean values for subgroups, 238–239*t*
extensors endurance, 223–224
external oblique, 102*t*
external rotators of hip assessment, 60

F
face validity, 7
fatigue, hop testing after, 166
figure 8 hop test, 160
firefighter, sample preemployment screening, 26*t*
fitness categories, according to percentages, 129*t*
flexed arm hang
    about, 242
    for boys, 255*t*
    for girls, 256*t*
flexibility, defined, 39
football
    1RM bench normative values, 172*t*
    1RM bench press percentile values, 168*t*
    1RM squat normative values, 169*t*
    1RM squat percentile values, 168*t*
    backward overhead medicine ball throw test, 247–248
    descriptive data, pro-agility times, 206*t*
    pro-agility times, 205*t*
    three-cone drill times, 207*t*
    T-test percentile ranks, 203*t*
    vertical jump normative values, 185*t*
    Wingate anaerobic power test descriptive data, 273*t*
Foreman, T., 156
forward lunge, 135, 136
four square step test, 106
front squat, normative values for elite weightlifters, 170*t*
Fujisawa, H., 107
functional ability tests, 159–163, 189*t*
functional medicine ball toss, Pearson correlations with isokinetic trunk rotation strength, 258*t*

functional movement screen, 23, 90, 276
functional reach test, 110
functional testing
    about, 4–5
    client preparation, 11–13
    common problems, 13–16
    essential components, 18*f*
    purpose, 9–10
    selection of tests, 10–11
    sequence of tests, 11, 132
    use of tests, 17–19
functional testing algorithm
    about, 5, 18
    after lower extremity injury, 20*f*
    basic preceding measurements, 19
    low back injury, 24*f*
functional throwing performance index, 254

G
gait, testing, 113
gastrocnemius, 84*t*
gastrocnemius assessment, 75
gastrosoleus, 84*t*
Gillespie, J., 252
girth assessment, 32
gluteus maximus, 79*t*, 81*t*, 100*t*
gluteus medius, 80*t*, 81*t*, 101*t*
gluteus minimus, 81*t*
gracilis, 101*t*
Gribble, P.A., 109
Guskiewicz, K.M., 114

H
Haennel, R.G., 248
hamstring length
    active supine 90/90 position assessment, 72, 83*t*
    passive supine 90/90 position assessment, 74, 83*t*
hamstring muscles, 82*t*
hamstring passive straight leg raise assessment, 73, 83*t*
handball, 1RM bench normative values, 172*t*
Harman, E., 10
Heegaard, J.H., 22
Hertel, J., 109
hexagon hop test, modified, 165
hexagon test (bilateral lower extremity jump), 164, 190*t*
hip abduction assessment, 88, 101*t*
hip adductor assessment (long *vs.* short), 67
hip adductor muscles, 82*t*
hip extension assessment, 87, 100–101*t*
hip extensor muscles, 79*t*
hip extensors assessment, 58
hip external rotator muscles, 81*t*

hip flexibility figure four test, 71
hip flexor muscles, 80*t*
hip flexors assessment, 59
hip internal rotator muscles, 81*t*
hockey, test descriptive data, 273*t*
hop-and-stop test
    about, 157
    normal clients' results, 188*t*
    predictive value, 188*t*
hop testing after fatigue, 166
hurdle step assessment, 93–94
hurdle test
    about, 198
    normative data, 208*t*

I
Ikeda, Y., 246, 249
iliacus, 80*t*
iliocostalis lumborum, 79*t*, 100*t*,
    102*t*
iliotibial band assessment, 68
Illinois agility test, 199
impairment, after injury, 4
inappropriate tests, 15–16
inferior gemellus, 81*t*
infraspinatus, 79*t*
injury
    anterior cruciate ligament injury,
      19–22
    considerations for return to com-
      petition, 11*t*
    impairment after, 4
    low back injury, 23, 24*f*, 25–26
    lower extremity injury, 20*f*, 21*t*
in-line lunge assessment, 95
insidious onset of neck pain, 212
interater reliability, 5
internal oblique, 102*t*
internal rotators of hip assessment,
    61
International Knee Documentation
    Committee, 149
intraclass correlation coefficient, 6
intraclass correlation coefficient
    benchmark values, 6*t*
intrarater reliability, 5
intratester reliability
    active hamstring 90/90 test, 83*t*
    passive hamstring 90/90 test, 83*t*
    straight leg raise test, 83*t*
isokinetic trunk rotation strength,
    Pearson correlations with func-
    tional medicine ball toss, 258*t*

J
Johnson, B.L., 105, 251
Jull, G., 212
Juris, P.M., 166

K
kappa benchmarks values, 6*t*
kappa statistic, 5, 90

Keenum,S., 252
kickboxing, test descriptive data,
    273*t*
Kier peak leg power nomogram,
    152
kinesthesia, testing, 103
knee bending in 30 seconds, 137
knee test protocol, 149
KT-1000 test, 19
Kuramoto, A.K., 145

L
laboratory setting, client tests in, 27
Ladeira, C.E., 218
Langford, G., 147
lateral flexors endurance (side bridge),
    221
lateral lunge, 135
latissimus dorsi, 78*t*
latissimus dorsi assessment, 46, 57
leaping, 157
left-right difference in functional abil-
    ity tests, 189*t*
Leger, L.A., 123
leg power nomogram, in watts, 152*f*
Lephart, S.M., 134
levator scapulae, 77*t*
lifting capacity assessment, 230–232
lifting tasks, 25–26
likelihood ratio interpretation, 8*t*
likelihood ratios, 7
limb symmetry index, 149–150
loaded forward reach, 230
longissimus thoracis, 79*t*, 100*t*, 102*t*
longus capitis, 100*t*, 233*t*
longus coli, 100*t*
longus colli, 233*t*
low back injury
    functional testing algorithm, 24*f*
    progression of client with, 23,
      25–26
low back pain, 211
low back testing, 25*t*
lower crossed syndrome, 39
lower extremity anaerobic power
    testing, 263
lower extremity assessments, 58–76
lower extremity functional reach
    test, 117*t*
lower extremity functional test, 18,
    22–23, 265, 271*t*
lower extremity injury
    exercises to enhance testing, 21*t*
    functional testing algorithm, 20*f*
lower extremity testing, 27
lower (sternal) pectoralis major
    assessment, 49
lower trunk dynamic stability assess-
    ment, 215–220
lumbar erector spinae, 79*t*

lumbar erector spinae assessment,
    54
lumbar radiculopathy, 23
lunge test, 135–136

M
Manske, R.C., 14, 46, 50
Margaria-Kalamen power test, 15,
    251
Markovic, G., 152
Mattacola, C.G., 136
Maulder, P., 268
maximal controlled leap, 157
Mayhew, J.L., 251
McArdle, W.D., 119
McCurdy, K., 147
McGill, S.M., 211, 224
measurement reliability, 5
measurement validity, 7
medicine ball toss
    about, 246
    backward overhead throw norma-
      tive data, 261*t*
    means, categorized by age, class
      level, and throw type, 206*t*
    normative values for elite weight-
      lifters, 206*t*
    scores, 259*t*
middle distance running, test descrip-
    tive data, 273*t*
middle scalene, 77*t*
modified hexagon hop test, 165
modified single-leg hop, onto force
    plate, 178*t*
modified unilateral squat strength,
    174*t*
movement analysis, 85
multifidus, 233*t*
multiple single-leg hop stabilization
    test, 115–116
multistage fitness test, 124
Munich, H., 139
muscle groups
    capital and cervical neck flexor
      muscles, 100*t*
    deep neck flexor muscles, 233*t*
    endurance ratios, 235*t*
    gastrosoleus muscles, 84*t*
    hamstring muscles, 82*t*
    hip adductor muscles, 82*t*
    hip extensor muscles, 79*t*
    hip external rotator muscles, 81*t*
    hip flexor muscles, 80*t*
    hip internal rotator muscles, 81*t*
    lumbar erector spinae muscles,
      79*t*
    neck muscles, 77*t*
    quadratus lumborum muscle, 79*t*
    suboccipital muscles, 77*t*
    upper arms muscles, 78*t*
    upper extremity muscles, 78–79*t*

muscle imbalances, 211
muscle length assessment, 27, 39–40
muscles
    for hip abduction analysis, 101t
    for hip extension analysis, 100–101t
    with predisposition for tightness, 39
    for trunk flexion analysis, 102t

N

neck muscles, 77t
neck pain, 211–212
negative predictive value, 8
Nelson, J.K., 105, 251
normative values
    1RM bench press, 172t
    1RM squat, 169t
    505 agility test, 207t
    abdominal stage test, 234t
    about, 40
    back squat, elite weightlifters, 170t
    backward overhead throw medicine ball toss, 261t
    body mass index, Canadian, 37t
    Canadian vertical jump, 183t
    closed kinetic chain upper extremity test, 262t
    endurance testing comparison, 235t
    hurdle test, 208t
    medicine ball toss, elite weightlifters, 206t
    relative bench press, 171t
    Rockport Walk Test, 128t
    slalom test, 208t
    T-test, 202t
    various selected tests, 189t
    vertical jump test, 183–184t, 185–186t
    waist-to-hip, 37t
Noyes, F.R., 149

O

Ober's test—iliotibial band assessment, 68
obliquus capitis inferior, 77t
obliquus capitis superior, 77t
obturator extremus, 81t
obturator internus, 81t
occupational setting, client testing, 23–26
Olmsted, L.C., 109
Onate, J.A., 114
O'Neal, Shaquille, 33
one-legged cyclic hop test, 138
one-mile run/walk
    boys' percentile scores, 127t
    girls' percentile scores, 126t

one-mile walk test, 120
overall accuracy, 8

P

Payne V.G., 145
Pearson coefficient benchmark values, 6t
Pearson product-moment correlation coefficient, 6
pectineus, 80t, 82t, 101t
pectoralis major assessment, 47
pectoralis minor assessment, 50
pectoralis minor assessment (Borstad method), 51
pelvic girth assessment, 35
performance testing. *See* functional testing
Petschnig, R., 138, 149, 155
Pfeifer, K., 138
physical function
    classical objective measures, 4–5
    testing of attributes, 18
piriformis, 80t, 81t, 101t
piriformis assessment, 69
piriformis FAIR test, 70
Plisky, P.J., 109
plyometric leap test, 269
positive predictive value, 8
posterior scalene, 77t
power, assessing, 23, 131
power testing, 263
practicality, of testing, 10–11
practice, prior to testing, 15
practice trials, 13
predictive values
    about, 7–8
    for hop-and-stop test, 188t
preemployment screening, client testing, 26–27
primary energy systems, major characteristics, 15t
pro agility (5-10-5) test
    about, 193
    descriptive data in various athletic populations, 206t
    percentile ranks among college athletes, 204t
    times for college football players in NFL combine, 205t
progressive isoinertial lifting evaluation, 231
prone bridge
    about, 228
    endurance mean values, symptomatic and asymptomatic clients, 239t
proprioception, testing, 22–23, 103
proprioceptive neuromuscular facilitation stretching, 12
psoas assessment I, 62

psoas assessment II, 63
psoas muscles, 80t, 101t, 102t
pull-up test, 243
push-up test, 245, 257t

Q

quadratus femoris, 81t
quadratus lumborum, 79t, 101t
quadratus lumborum assessment, 55
quadratus lumborum assessment II, 56
quickness, 191

R

rectus abdominis, 102t
rectus capitis anterior, 100t, 233t
rectus capitis lateralis, 100t, 233t
rectus capitis posterior major, 77t
rectus capitis posterior minor, 77t
rectus femoris, 80t
rectus femoris assessment, 64–65
Reid, A., 155
relative bench press, normative values, 171t
relevance, of testing, 10
reliability measures, value descriptions, 6t
repetitive box-lifting task, 25, 232
research, utilizing carioca test, 167t
rest, between test repetitions, 14
rhomboid major, 78t
rhomboid minor, 78t
Riemann, B.L., 114
Riseberg, M.A., 158, 162
risk assessment, prior to testing, 13
Roberts, A., 122
Robinson, R.H., 109
Rockport Walk Test
    about, 121
    normative values, 128t
    sample use, 23
Roetert, E.P., 246
Romberg test, 111
rotational stability assessment, 99
Rousch, J.R., 253
Rudolf, M.C., 32
rugby, 1RM bench normative values, 172t
running-based anaerobic sprint test, 264

S

safety, of testing, 10
Salonia, M.A., 246
sartorius, 80t
seated chest pass, 250
seated shot-put throw
    about, 251–252
    correlations to, 262t
    physical performance characteristics, 261t

segmental multifidus test, 215
Sekiya, I., 150
Sell, K.E., 149
semimembranosus, 79t, 82t, 101t
semitendinosus, 79t, 82t, 100t
sensitivity, 7
serratus anterior, 78t
shoulder mobility assessment, 96
shuttle run
    about, 264
    percentile ranks in athletes, 271t
sidearm medicine ball throw
    about, 249
    correlations to selected variables, 257t
    distance, 257t
side bridge (lateral flexors endurance), 221
side-hop test, 162
side-step test, 107
single-hop test, 163
single-leg crossover hop for distance test
    about, 156
    correlations with isokinetic tests, 188t
    studies, scores, and reliability, 187t
single-leg hop
    about, 21–22
    correlations to various parameters, 179t
    onto force plate, modified, 178t
single-leg hop for distance
    about, 149–150
    studies, scores, and reliability, 176–177t
single-leg inclined squat test, 139
single-leg squat, 146, 147
single-leg stance test, 104
Sinnett, A.M., 269, 270
sit-up endurance test
    about, 220
    interpretation, 234t
six-meter timed hop, 154
slalom test
    about, 197
    normative data, 208t
slow static stretching, 12
SnNout, 7
soccer
    1RM bench normative values, 172t
    1RM squat normative values, 169t
    abdominal stage test normative values, 234t
    descriptive data, pro-agility times, 206t

T-test descriptive data, 204t
T-test percentile ranks, 203t
vertical jump normative values, 186t
Wingate anaerobic power test descriptive data, 273t
softball
    Underkoffler softball throw for distance, 245
    Wingate anaerobic power test descriptive data, 273t
soleus, 84t
soleus muscle length test, supine, 76
specificity, 7, 10
speed, 191
speed skating, test descriptive data, 273t
spinal instability, 211
spinal stenosis, 211
SpPin, 7
sprinting (competitive), test descriptive data, 273t
squat jump test
    about, 268
    intraclass correlation with other tests, 274t
    mean and symmetry index, 274t
    statistical analyses, 274t
squat strength, modified unilateral, 174t
stair hopple test, 158
standing long jump
    about, 148
    percentile ranks in 15- and 16-year-old athletes, 175t
    percentile ranks in elite athletes, 175t
    test values, 175t
Star Excursion Balance Test
    about, 108–109
    client with low back injury, 25
    of college baseball team, 22–23
    dynamic function assessment, 19
    and specificity of training, 15
    test reliability, 117t
static balance, stork test, 105
static stretching, 12
statistical terminology, 5–8
statistics
    1RM leg press, 143
    20-meter shuttle run test, 123
    300-meter sprint, 270
    abdominal endurance test, 224, 226
    balance error scoring system, 114
    bilateral lower extremity jump, 164
    body mass index assessment, 33
    Bosco test, 267

carioca test, 134
closed kinetic chain upper extremity test, 253
co-contraction test, 132
deep neck flexor test, 213–214
double-leg lowering test, 218
extensors assessment, 58, 223, 227
external rotators of hip assessment, 60
figure 8 hop test, 160
four square step test, 106
functional reach test, 110
functional throwing performance index, 254
gastrocnemius assessment, 75
girth assessment, 32
hamstring assessment, 72, 73, 74
hexagon hop test, modified, 165
hexagon test (bilateral lower extremity jump), 164
hip flexors assessment, 59
hop testing after fatigue, 166
hurdle test, 198
iliotibial band assessment, 68
internal rotators of hip assessment, 61
lateral flexors endurance (side bridge), 221
latissimus dorsi assessment, 46
loaded forward reach, 230
lower extremity functional test, 265
lunge test, 136
maximal controlled leap, 157
medicine ball toss, 246, 248, 249
modified hexagon hop test, 165
multiple single-leg hop stabilization test, 116
multistage fitness test, 124
Ober's test—iliotibial band assessment, 68
one-legged cyclic hop test, 138
pectoralis assessment, 47, 48, 49, 50, 51
pelvic girth assessment, 35
piriformis FAIR test, 70
progressive isoinertial lifting evaluation, 231
prone bridge, 228
repetitive box-lifting task, 232
Rockport Walk Test, 121
running-based anaerobic sprint test, 264
seated shot-put throw, 251–252
side-hop test, 162
side-step test, 107
single-leg crossover hop for distance test, 156

statistics *(continued)*
　　single-leg hop for distance, 149–150
　　single-leg stance test, 104
　　six-meter timed hop, 154
　　slalom test, 197
　　soleus muscle length test, supine, 76
　　squat tests, 92, 139, 142, 268
　　stair hopple test, 158
　　standing long jump, 148
　　Star Excursion Balance Test, 108–109
　　step-down, 133
　　stork test, 105
　　supine bridge, 229
　　torso height assessment, 34
　　triple hop for distance, 155
　　triple jump for distance test, 153
　　T-test, 192
　　up-down test, 161
　　vertical jump test, 152
　　waist-to-hip ratio assessment, 33
　　Wingate anaerobic power test, 266
　　zigzag run test, 201
statistics test procedure, plyometric leap, 269
step-down test
　　about, 133
　　reliability, 167*t*
sternocleidomastoid, 77*t*, 100*t*
sternocleidomastoid assessment, 44
Stockbrugger, B.A., 248
stork test, 105
strength, assessing, 23, 131
stretching
　　comparison of techniques, 12*t*
　　effects of, 39
suboccipital muscles, 77*t*
suboccipital muscles assessment, 45
superior gemellus, 81*t*
supine bridge
　　about, 229
　　endurance mean values, symptomatic and asymptomatic clients, 239*t*
supraspinous, 78*t*
Sykes, K., 122

T
Tabor, M.A., 265
Takeda, R., 107
tandem walking, 112
Tegner, Y., 149
tennis, test descriptive data, 273*t*
tennis players, backward movement agility test data, 208*t*
tensor fasciae latae, 80*t*, 81*t*, 101*t*
tensor fasciae latae assessment, 66
teres major, 79*t*

teres minor, 79*t*
test analysis and interpretation
　　1.5 mile run test, 125
　　1RM bench press, 144
　　1RM leg press, 143
　　12-minute run test, 125
　　20-meter shuttle run test, 123
　　300 shuttle run, 264
　　505 agility test, 194
　　abdominal endurance test, 224, 226
　　abdominal stage test, 219
　　active straight leg raise assessment, 97
　　alternate pull-up test, 244
　　backward movement agility test, 200
　　balance error scoring system, 114
　　bilateral lower extremity jump, 164
　　body mass index assessment, 33
　　Bosco test, 267
　　carioca test, 134
　　cervical deep flexor muscle assessment, 86
　　Chester step test, 122
　　closed kinetic chain upper extremity test, 253
　　co-contraction test, 132
　　deep neck flexor test, 213
　　double-leg lowering test, 218
　　Edgren side-step test, 195
　　extensors assessment, 58, 223, 227
　　external rotators of hip assessment, 60
　　figure 8 hop test, 189*t*
　　flexed arm hang, 242
　　four square step test, 106
　　functional throwing performance index, 254
　　girth assessment, 32
　　hamstring assessment, 73, 74
　　hexagon hop test, modified, 165
　　hexagon test (bilateral lower extremity jump), 164
　　hip abduction assessment, 88
　　hip adductor assessment (long *vs.* short), 67
　　hip extension assessment, 87
　　hip flexibility figure four test, 71
　　hip flexors assessment, 59
　　hop testing after fatigue, 166
　　hurdle step assessment, 93
　　hurdle test, 198
　　iliotibial band assessment, 68
　　Illinois agility test, 208*t*
　　in-line lunge assessment, 95
　　internal rotators of hip assessment, 61

knee bending in 30 seconds, 137
lateral flexors endurance (side bridge), 221
latissimus dorsi assessment, 46, 57
loaded forward reach, 230
lower extremity functional test, 265
lumbar erector spinae assessment, 54
lunge test, 135–136
maximal controlled leap, 157
medicine ball toss, 246, 247–248, 257*t*
modified hexagon hop test, 165
multiple single-leg hop stabilization test, 115–116
multistage fitness test, 124
Ober's test—iliotibial band assessment, 68
one-legged cyclic hop test, 138
one-mile walk test, 120
pectoralis assessment, 47, 50
pelvic girth assessment, 35
piriformis assessment, 69, 70
pro agility (5-10-5) test, 193
progressive isoinertial lifting evaluation, 231
prone bridge, 228
psoas assessment, 62, 63
pull-up test, 243
push-up test, 245
quadratus lumborum assessment, 55
quadratus lumborum assessment II, 56
rectus femoris assessment, 65
Rockport Walk Test, 121
Romberg test, 111
rotational stability assessment, 99
running-based anaerobic sprint test, 264
seated chest pass, 250
seated shot-put throw, 251
segmental multifidus test, 215
shoulder mobility assessment, 96
side-hop test, 189*t*
side-step test, 107
single-hop test, 189*t*
single-leg crossover hop for distance test, 156
single-leg hop for distance, 149
single-leg stance test, 104
sit-up endurance testing, 220
six-meter timed hop, 154
slalom test, 197
squat tests, 91, 139, 142, 146, 147, 268
stair hopple test, 158

standing long jump, 148
Star Excursion Balance Test, 108
step-down, 133
sternocleidomastoid assessment, 44
stork test, 105
suboccipital muscles assessment, 45
supine bridge, 229
tandem walking, 112
tensor fasciae latae assessment, 66
three-cone drill test, 196
Tinetti test, 113
torso height assessment, 34
triple hop for distance, 155
triple jump for distance test, 153
trunk curl-up test, 216
trunk flexion assessment, 89
trunk stability push-up assessment, 98
T-test, 192
Underkoffler softball throw for distance, 245
up-down test, 189*t*
upper extremity assessments, 41
upper trapezius assessment, sitting, 42
upper trapezius assessment, supine, 43
vertical jump test, 151–152
waist-to-hip ratio assessment, 33
Wingate anaerobic power test, 266
zigzag run test, 201
testing, procedures and rules, 13–14
test procedure
    1.5 mile run test, 125
    1RM bench press, 144
    1RM leg press, 143
    12-minute run test, 125
    20-meter shuttle run test, 123
    300-meter sprint, 270
    300 shuttle run, 264
    505 agility test, 194
    abdominal endurance test, 224, 226
    abdominal stage test, 219
    active straight leg raise assessment, 97
    alternate pull-up test, 244
    backward movement agility test, 200
    balance error scoring system, 114
    biceps assessment, 53
    bilateral lower extremity jump, 164
    body mass index assessment, 33
    Bosco test, 267
    carioca test, 134

cervical deep flexor muscle assessment, 86
Chester step test, 122
closed kinetic chain upper extremity test, 253
co-contraction test, 132
deep neck flexor test, 213
double-leg lowering test, 217–218
Edgren side-step test, 195
extensors assessment, 227
extensors endurance, 223
external rotators of hip assessment, 60
figure 8 hop test, 160
flexed arm hang, 242
four square step test, 106
functional reach test, 110
functional throwing performance index, 254
gastrocnemius assessment, 75
girth assessment, 32
hamstring assessment, 72, 73, 74
hexagon hop test, modified, 165
hexagon test (bilateral lower extremity jump), 164
hip abduction assessment, 88
hip adductor assessment (long *vs.* short), 67
hip extension assessment, 87
hip extensors assessment, 58
hip flexibility figure four test, 71
hip flexors assessment, 59
hop testing after fatigue, 166
hurdle step assessment, 93
hurdle test, 198
iliotibial band assessment, 68
in-line lunge assessment, 95
internal rotators of hip assessment, 61
knee bending in 30 seconds, 137
lateral flexors endurance (side bridge), 221
latissimus dorsi assessment, 46, 57
loaded forward reach, 230
lower extremity functional test, 265
lumbar erector spinae assessment, 54
lunge test, 135
maximal controlled leap, 157
medicine ball toss, 246, 247, 249
modified hexagon hop test, 165
multiple single-leg hop stabilization test, 115
multistage fitness test, 124
Ober's test—iliotibial band assessment, 68
one-legged cyclic hop test, 138

one-mile walk test, 120
pectoralis assessment, 47, 48, 49, 50, 51
pelvic girth assessment, 35
piriformis assessment, 69, 70
plyometric leap test, 269
pro agility (5-10-5) test, 193
progressive isoinertial lifting evaluation, 231
prone bridge, 228
protocol for 1RM testing, 104
psoas assessment, 62, 63
pull-up test, 243
push-up test, 245
quadratus lumborum assessment, 55
quadratus lumborum assessment II, 56
rectus femoris assessment, 64
repetitive box-lifting task, 232
Rockport Walk Test, 121
Romberg test, 111
rotational stability assessment, 99
running-based anaerobic sprint test, 264
seated chest pass, 250
seated shot-put throw, 251
segmental multifidus test, 215
shoulder mobility assessment, 96
side-hop test, 162
side-step test, 107
single-hop test, 163
single-leg crossover hop for distance test, 156
single-leg hop for distance, 149
single-leg stance test, 104
sit-up endurance testing, 220
six-meter timed hop, 154
slalom test, 197
soleus muscle length test, supine, 76
squat tests, 91–92, 139, 141–142, 146, 147, 268
stair hopple test, 158
standing long jump, 148
Star Excursion Balance Test, 108
step-down, 133
sternocleidomastoid assessment, 44
stork test, 105
suboccipital muscles assessment, 45
supine bridge, 229
tandem walking, 112
tensor fasciae latae assessment, 66
three-cone drill test, 196
Tinetti test, 113
torso height assessment, 34
triceps assessment, 52
triple hop for distance, 155

test procedure *(continued)*
   triple jump for distance test, 153
   trunk curl-up test, 216
   trunk flexion assessment, 89
   trunk stability push-up assessment, 98
   T-test, 192
   Underkoffler softball throw for distance, 245
   up-down test, 161
   upper extremity assessments, 41
   upper trapezius assessment, sitting, 42
   upper trapezius assessment, supine, 43
   vertical jump test, 151–152
   waist-to-hip ratio assessment, 33
   Wingate anaerobic power test, 266
   zigzag run test, 201
three-cone drill test
   about, 196
   times for college football players in NFL combine, 207t
Tinetti test
   about, 113
   Assessment Tool for Balance form, 277
   Assessment Tool for Gait form, 278
torso height assessment, 34
transverse abdominis, 102t
trapezius, 77t
triceps assessment, 52
triceps brachii, 78t
triple hop for distance, 155
triple jump for distance test, 153
trunk assessments, 54–57, 211–212
trunk curl-up test, 216
trunk endurance assessment, 25
trunk flexion assessment, 89, 102t
trunk muscle endurance assessment, 221–229, 230–232
trunk stability push-up assessment, 98
T-test
   about, 23, 192
   deciles for female and male college-age subjects, 202t

descriptive data in various athletic populations, 204t
high-school soccer players percentile ranks, 203t
mean scores for female and male college-age subjects, 202t
normative values for athletes, 202t
percentile ranks of college football players, 203t

U
Underkoffler softball throw for distance, 245
untrained testers, 15
unwanted compensation patterns, 14
up-down test, 161
upper arms muscles, 78t
upper crossed syndrome, 39
upper extremity assessments, 41–53, 241
upper extremity muscles, 78–79t
upper extremity throwing analysis form, 281–283
upper (clavicular) pectoralis major assessment, 48
upper trapezius assessment
   sitting, 42
   supine, 43

V
validity, 7, 10
van der Harst, J.J., 22
Vertec jumping apparatus, 23, 151
vertical jump test
   about, 151–152
   basketball players percentile ranks, 181t
   deciles for female and male college-age subjects, 184t
   NCAA female athlete percentile values, 182t
   normative data by gender and age, Canadians, 183–184t
   normative data for male and female Canadians, 183–184t
   normative values for athletes in various sports, 185–186t
   volleyball players percentile ranks, 182t

youth percentile ranks for heights, 180t
volleyball
   1RM bench normative values, 172t
   1RM squat normative values, 169t
   descriptive data, pro-agility times, 206t
   player vertical jump percentile rates, 182t
   T-test descriptive data, 204t
   vertical jump normative values, 185t
$\dot{V}O_2$max
   cardiorespiratory fitness classification, 128t
   defined, 119
   determining, 120, 121

W
waist-to-hip normative values, 37t
waist-to-hip ratio assessment, 33
warm-up, before testing, 12–13, 132
web site, functional movement, 92
weightlifting
   front and back squat normative values, 170t
   vertical jump normative values, 186t
whiplash-associated disorder, 212
Whitney, S.L., 106
Wilk, K.E., 154, 156
Wingate anaerobic power test
   about, 266
   descriptive data in athletic populations, 272–273t
   percentile ranks for physically active college-aged males and females, 272t
Wingate bike test, 15
wrestling, vertical jump normative values, 186t

Y
Youdas, J.W.., 74
yo-yo test, 124

Z
zigzag run test, 201
Zillmer, D.A., 5, 18

About the Authors

Michael P. Reiman, PT, DPT, MEd, OCS, ATC, CSCS, is an assistant professor of physical therapy at Wichita State University. As a clinician certified in functional movement screen testing, Reiman has over 18 years of experience in assessing, rehabilitating, and training athletes, clients, and patients at various levels of ability. He received his doctoral degree in physical therapy from MGH Institute of Health Professions in 2007. Previously, Reiman was owner and operator of a business specializing in personal strength and conditioning for athletes. In addition to his certifications as a board-certified orthopedic physical therapist, an athletic trainer, and strength and conditioning specialist, Reiman is a USA Weightlifting level 1 coach and a USA Track and Field level 1 coach.

Reiman has published two book chapters on training for strength, power, and endurance as well as multiple articles in such journals as the *American Journal of Sports Medicine*, *Journal of Orthopedic and Sports Physical Therapy*, and *Journal of Sport Rehabilitation*. He presents on various levels of assessment and treatment methods at national, regional, and local conferences and actively participates in research regarding various testing methods for performance. His current research and presentation interests focus on performance enhancement, low back pain, and trunk endurance in athletes and individuals in occupational environments. He is currently completing a manual therapy fellowship from the American Academy of Orthopaedic Manual Physical Therapists and continues to practice clinically on various orthopedic and sport-related injuries. Reiman currently serves on the editorial board for the *Journal of Sport Rehabilitation* and is a reviewer for the *Journal of Orthopaedic and Sports Rehabilitation* and the *Journal of Manual and Manipulative Therapy*.

In 2007, Reiman was the recipient of the Rodenberg Excellence in Teaching Award from the College of Health Professions at Wichita State University. He is a member of the American Physical Therapy Association, National Athletic Trainers' Association, National Strength and Conditioning Association, USA Weightlifting Association, and USA Track and Field Association.

In his free time, Reiman enjoys spending time with his family, coaching youth sports, watching college football, and bicycling. He resides in Colwich, Kansas.

Robert C. Manske, PT, DPT, MEd, SCS, ATC, CSCS, is an associate professor of physical therapy at Wichita State University. He earned a doctoral degree in physical therapy in 2006 from the MGH Institute of Health Professions. Manske was also a sport physical therapy fellow, training under the guidance of George J. Davies in one of the first sport physical therapy residency programs. As a practicing physical therapist, Manske has over 14 years of clinical experience in rehabilitation and is currently researching knee and shoulder rehabilitation and sport performance enhancement.

Manske has published multiple book chapters, research studies, articles, and home study courses regarding sport rehabilitation and presented his research at international, national, regional, and local rehabilitation continuing education venues. Manske is the editor of *Postsurgical Orthopedic Sports*

Rehabilitation: Knee and Shoulder and an APTA Sports Section monograph titled *Patellofemoral Joint Revisited: Implications for the 21st Century.* He is a board-certified sport physical therapist, certified athletic trainer, and certified strength and conditioning specialist. He is also a member of American Physical Therapy Association and the National Athletic Trainers' Association. He serves as chair of the Knee Special Interest Group and chair of the membership committee for the Sports Section of the APTA. Manske presents 15 to 20 weekend courses on various shoulder and knee topics throughout the year and still remains active in clinical practice using functional testing on multiple orthopedic and athletic patients weekly.

In 2007, Manske received the Sports Section Excellence in Education Award from the American Physical Therapy Association. He has also received the Kansas Physical Therapy Educator award from the Kansas Physical Therapy Association (2003) and the Rodenberg Teaching Award from the College of Health Professions at Wichita State University (2004).

Manske and his wife, Julie, live in Wichita. He enjoys spending time with his family and watching college basketball.

DVD Menu
and User Instructions

DVD Menu

Fundamental Movement Testing

Hip Extension Assessment
Hip Abduction Assessment
Trunk Flexion Assessment
Hurdle Step Assessment
In-Line Lunge Assessment
Shoulder Mobility Assessment
Active Straight Leg Raise Assessment
Rotational Stability Assessment

Balance

Star Excursion Balance Test
Functional Reach Test
Multiple Single-Leg Hop Stabilization Test

Strength and Power

Single-Leg Inclined Squat Test
Alternate Single-Leg Squat Test
Vertical Jump Test
Triple Jump for Distance Test
Single-Leg Crossover Hop for Distance Test
Figure 8 Hop Test
Hexagon Test (Bilateral Lower Extremity Jump)
Modified Hexagon Hop Test

Speed, Agility, and Quickness Testing

T-Test
Pro Agility (5-10-5) Test
505 Agility Test
Edgren Side Step Test
Three-Cone Drill Test
Illinois Agility Test
Zigzag Run Test

Trunk Testing

Deep Neck Flexor Test
Segmental Multifidus Testing
Trunk Curl-Up Test
Double-Leg Lowering Test
Abdominal Stage Test
Endurance of Abdominal Muscles Test
Extensor Dynamic Endurance Test

Upper Extremity Testing

Alternate Pull-Up Test
Backward Overhead Medicine Ball Throw Test
Closed Kinetic Chain Upper Extremity Test

Lower Extremity Anaerobic Power

Lower Extremity Functional Test
Squat Jump Test
Plyometric Leap Test

DVD User Instructions

The reproducible forms on this DVD-ROM can only be accessed using a DVD-ROM drive in a computer (not a DVD player on a television). To access the reproducible forms, follow these instructions:

Microsoft Windows

1. Place DVD in the DVD-ROM drive of your computer.
2. Double-click on the "My Computer" icon from your desktop.
3. Right-click on the DVD-ROM drive and select the "Open" option from the pop-up menu.
4. Double-click on the "Reproducible Forms" folder.
5. Select the reproducible form that you want to view or print.

Macintosh

1. Place DVD in the DVD-ROM drive of your computer.
2. Double-click the DVD icon on your desktop.
3. Double click on the "Reproducible Forms" folder.
4. Select the reproducible form that you want to view or print.

Note: You must have Adobe Acrobat reader to view the reproducible forms.